CRIME SCIENCE

CRIME
SCIENCE

METHODS
OF FORENSIC
DETECTION

JOE NICKELL AND
JOHN F. FISCHER

THE UNIVERSITY PRESS OF KENTUCKY

Copyright © 1999 by The University Press of Kentucky

Scholarly publisher for the Commonwealth,
serving Bellarmine University, Berea College, Centre
College of Kentucky, Eastern Kentucky University,
The Filson Historical Society, Georgetown College,
Kentucky Historical Society, Kentucky State University,
Morehead State University, Transylvania University,
University of Kentucky, University of Louisville,
and Western Kentucky University.
All rights reserved.

Editorial and Sales Offices: The University Press of Kentucky
663 South Limestone Street, Lexington, Kentucky 40508-4008
www.kentuckypress.com

08 07 06 05 04   10 9 8 7 6

Library of Congress Cataloging-in-Publication Data
Nickell, Joe.
   Crime science : methods of forensic detection / Joe Nickell
and John F. Fischer
      p.   cm.
   Includes bibliographical references and index.
   ISBN 0-8131-2091-8 (cloth : alk. paper)
   1. Criminal investigation. 2. Forensic sciences. 3. Criminal inves-
tigation—Technological innovations. I. Fischer, John F. II. Title.
HV8073.N517   1998
365.25—dc21   98–30749

This book is printed on acid-free recycled paper meeting
the requirements of the American National Standard
for Permanence in Paper for Printed Library Materials.

Manufactured in the United States of America

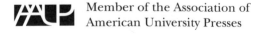 Member of the Association of
American University Presses

CONTENTS

ACKNOWLEDGMENTS

The authors are grateful to the following individuals for special assistance: Bill Schulz, Orlando Police Department, Orlando, Florida; Robert Miller, Phoenix Police Department, Phoenix, Arizona; Robert Ruttman, Mesa Police Department, Mesa, Arizona; and Marvin Stephens, Florida Department of Law Enforcement, Tallahassee, Florida.

The authors are also grateful to the following agencies for helpful cooperation: Orange County Sheriff's Office, Orlando, Florida; Mesa Police Department, Mesa, Arizona; Phoenix Police Department, Phoenix, Arizona; and the Orlando Police Department, Orlando, Florida.

1 INTRODUCTION TO FORENSIC SCIENCE AND CRIMINALISTICS

The rational basis upon which the work of today's investigator is predicated is called the scientific method. This method is *empirical* (from the Latin *empiricus,* "experienced"), meaning that knowledge is gained from direct observation. Underlying the empirical attitude is a belief that there is a real knowable world that operates according to fixed rules and that effects do not occur without causes.[1] In contrast to, say, religious dogma, science is open-ended: It is amenable to being amplified or to having its errors corrected in the light of new evidence. For our purposes, the scientific method is one that involves *analysis* (studying the unknown item to determine its essential characteristics), *comparison* (examining how the characteristics compare with the established properties of known items), and *evaluation* (assessing the similarities and dissimilarities for identification purposes).[2]

Forensic means characteristic of, or suitable for, a court of law. Hence *forensic science* is a broad term that embraces all of the scientific disciplines that are utilized in investigations with the goal of bringing criminals to justice. The American Academy of Forensic Sciences defines it as "the study and practice of the application of science to the purposes of the law."[3] It includes such fields as forensic medicine, toxicology, psychology, and anthropology as well as the work of specialized examiners of fingerprints, firearms, tool marks, and questioned documents. The term is so broad as to include even criminology, a social science that plays a role in the administration of civil law.

1

Like criminology, *criminalistics* is a division of forensic science, and its practitioner is called a "criminalist." The discipline has been variously defined. In 1963, the California Association of Criminalistics adopted the following definition: "Criminalistics is that profession and scientific discipline directed to the recognition, identification, individualization and evaluation of physical evidence by application of the natural sciences to law-science matters."[4] This definition has since been adopted by the American Academy of Forensic Sciences.

At one time a criminalist might have been a generalist, but now most are specialists in one of many areas of expertise. These specialties include forensic chemistry, toxicology, drug analysis, serology, trace analysis, latent fingerprints, firearms examination, impression evidence, questioned document examination, and voice examination. Depending on the crime laboratory, some specialties may be combined with or separated from others. For example, because of the way they are compared, tool marks may be handled by the firearms examiner. Conversely, because of the increasing complexity of DNA analysis, it may be treated as a discipline separate from forensic serology, and crime scene examination may itself now be a specialized area of expertise.

In certain of the criminalistics disciplines, such as firearms examination and fingerprint identification, there is occasionally confusion between the terms *identify* and *individualize*. Although their respective Latin roots are similar (*idem* means "the same" and *individus* means "not divisible"), the terms are quite distinctive as they are correctly applied in criminalistics. As the great forensic authority Paul L. Kirk explains: "The terms 'identification' and 'identity' are used constantly by practitioners in the field. Few stop to define the terms. 'Identity' is defined by all philosophical authorities as uniqueness. . . . Bowing to general scientific usage, we must however accept the term 'identification' in a broader context referring only to placing the object in a restricted class. . . . In this sense, the criminalist would identify the object as a paint chip, but not relate it to the painted surface from which the chip was derived. He would identify a marking as a fingerprint, but without relation to the hand that placed it." Kirk raises an interesting point, noting that "for the criminalist to use the word 'identification' in its accepted context is to admit that there is no reason for his special existence. . . . The criminalist does not attempt identification except as a prelude to his real function—that of individualizing. The real aim of all forensic science is to establish individuality, or to approach it as closely as the present state of the science allows. Criminalistics is the science of individualization."[5]

Individualization—that is, demonstrating the uniqueness of some item of evidence—is made possible by the fact that no two things in nature are *exactly* the same. Nearly everyone knows that this is true of snowflakes and fingerprints. But it is also true of gun barrels, lip impressions, shoe prints, and pieces of broken glass. The principle that *all objects in the universe are unique* may be expressed in many ways:

> No two things that happen by chance ever happen in exactly the same way.
> No two things are ever constructed or manufactured in exactly the same way.
> No two things ever wear in exactly the same way.
> No two things ever break in exactly the same way.[6]

Of course, items made from a mold will be very similar to each other, and as a practical matter it may not be possible to individualize, say, the track of a brand new tire. If the impression was clear, however, it should be possible to identify its *class characteristics,* discovering which brand and model of tire the tread pattern came from. Over time, the tire acquires nicks, patterns of wear, and sufficient *individual characteristics,* making individualization possible. In the case of a new tread, one could only say that the track *could* have come from a certain automobile—that the track was *consistent* with it. In the case of a nicked and worn tire, the criminalist would be able to compare the *questioned* evidence (the tread imprint) with that of a *known standard* (the suspect's tire or a test imprint made from it) to see whether or not the questioned impression may be individualized.

How many similarities are required to individualize an impression? In his book *Individualization: Principles and Procedures in Criminalistics,* Harold Tuthill writes, "The individualization of an impression is established by finding agreement of corresponding individual characteristics by such number and significance as to preclude the possibility (or probability) of their having occurred by mere coincidence, and establishing that there are no differences that cannot be accounted for." Tuthill notes that the words "an impression" may be replaced by "a bullet," "handwriting," "a break or fracture," or other wording appropriate to the evidence under consideration.[7] So, there is no hard and fast rule; the number of similarities required for individualization will depend on the unique quality of the details discovered. "Whether or not we have found sufficient [numbers] will be decided, subjectively, in each case," Tuthill observes. "One will suffice, if it can be seen to be unique. It is for this reason that it is important for the criminalist to be well trained and have considerable experience before undertaking to give evidence as an expert witness."[8]

In the process of individualization, the scientific examiner sometimes makes use of probability and may cite Newcomb's Rule (after Professor Simon Newcomb): "The probability of concurrence of all events is equal to the continued product of the probabilities of all the separate events." Consider a document typed on an old fashioned typewriter with the standard forty-two type bars and eighty-four characters. Suppose that examination of the document reveals that the machine that typed it has five individual characteristics—one caused by a misaligned type bar (a $1/42$ probability) and the others caused by defects in individual characters (or $1/84$ probability each). The mathematical formula for this set of circumstances would be: $1/42 \times 1/84 \times 1/84 \times 1/84 \times 1/84 = 1/2,091,059,712$— a figure greater than the total number of typewriters of that particular model in existence.[9] In many cases of individualization, the probability of each factor or "event" is unknown, and courts tend to take a dim view of assigning arbitrary values. Nevertheless, individualization, after the common-sense approach of multiplying each unusual factor by each additional one has been explained, represents the basis of many forensic comparisons presented in court.

The evidence that is encountered by the forensic scientist may be classified in various ways. Perhaps the simplest is the division into *personal evidence,* in the form of personal testimony, such as that of an eyewitness, and *physical evidence,* such as fingerprints or glass fragments. Personal evidence is subjective and colored by the person's attitudes and perceptions. In contrast, physical evidence is objective and remains the same for each observer, although there may be subjective aspects.[10] Another system of classifying evidence is by the type of crime. Yet another is by the general nature of evidence: *biological, physical,* and *miscellaneous* (including polygraph tests, voice analysis, and photography). Still other methods are employed, including classification by the most-appropriate-laboratory approach. "In this scheme," states one forensic text, "evidence is placed into categories according to whether it is simply to be identified, an individualization is sought, or a reconstruction of the event is desired. Looked at in this way, evidence is examined and analyzed in a manner that is relevant to the investigation."[11] In this regard we should mention *corpus delicti* evidence, the Latin phrase meaning "the body [or the essential elements] of the crime." This would generally be, for example, the body of a victim in a homicide or a broken or open strongbox in a burglary.

After results of a forensic analysis have been obtained, they must often be communicated to nonspecialists, including jurors. Whereas, generally speaking, a witness can testify only to facts that he or she knows, an expert witness is employed "because the issues require analysis and explanation by a person with scientific or specialized knowledge or experience."[12] Hence an expert may give an opinion to the court when it is relevant both to the facts of the case and to the analysis conducted. The professional criminalist should withstand any pressure to support the case of the attorney who enlists him. Scientific integrity demands that he be as willing to clear the innocent as to incriminate the guilty.

The landmark case for the admissibility of scientific procedures and their results is *Frye v. United States* (1923). In *Frye,* the court stated: "Just when a scientific principle or discovery crosses the line between the experimental and demonstrable stages is difficult to define. Somewhere in this twilight zone the evidential force of the principle must be recognized, and while courts will go a long way in admitting expert testimony deduced from a well-recognized scientific principle or discovery, the thing from which the deduction is made must be sufficiently established to have gained general acceptance in the particular field in which it belongs." Decisions subsequent to Frye made clear that *general* recognition—familiarity with a test or procedure by every scientist in the field—is not required for admissibility. Also "the particular field in which it belongs" was narrowed to mean the applicable specialty or subspecialty within a scientific field. And, in some instances, results or experiments devised in light of a particular problem were admitted if they were based on accepted principles of analysis and if a proper foundation has been laid.[13]

In 1993, however, the U.S. Supreme Court, in *Daubert v. Merrell Dow Pharmaceuticals, Inc.,* rejected the *Frye* "general acceptance" rule. An article in *Journal of Forensic Sciences* noted that many state high courts "have a very good grasp of scientific evidence and have enunciated readily-applied rules by which their trial courts are to evaluate it" and lamented the fact that "junk science" might find its way into the courtroom.[14] The Supreme Court rejected these concerns, stating, "Vigorous cross-examination, presentation of contrary evidence, and careful instruction on the burden of proof are the traditional and appropriate means of attacking shaky but admissible evidence." The Court offered certain guidelines for gauging the validity of scientific evidence, emphasizing flexibility: the technique or theory must be testable and must have in fact been tested; it must have been subjected to peer review and the publication process; standards must exist and be maintained that control the operation of the technique; and the method or theory must have been widely accepted within the relevant scientific discipline.[15]

HISTORY

While it is true that, as one forensic expert notes, "the idea of using science as an aid in criminal investigation was foreshadowed in the fictional works of Sir Arthur Conan Doyle at the turn of the century,"[16] there were in fact scientific advancements in crime detection long before Sherlock Holmes made his debut in 1887.

The earliest forensic scientists were medical men, who logically happened to be among the first on the scene of a death. The earliest record of physicians applying medical knowledge to the solving of crimes is the Chinese book *Hsi Duan Yu* ("The Washing Away of Wrongs") dating from 1248. Though it contains many unscientific notions, it also offers some important medical-legal procedures, such as distinguishing a case of drowning by the presence of water in the victim's lungs and identifying a death from strangulation by observing the characteristic pressure marks on the throat and damaged cartilage in the neck.[17]

There were few advancements until the eighteenth century, when French medical jurist Antoine Louis sought not only to identify the cause of death but also to distinguish between murder and suicide in questionable cases and to solve other relevant matters.[18] Eventually, such pathologists saw the utility of attempting to establish the victim's identity, determine the time of death, identify poisons, analyze bloodstains, and seek other advancements in the field of medical jurisprudence.

The end of the eighteenth century saw the origins of modern chemistry, which paved the way for the science of toxicology. Not surprisingly, the man considered "the father of forensic toxicology"[19] was a medical teacher named Mathieu Orfila (1787-1853). Shortly after his graduation from the university in 1811, the Spanish-born Orfila became a private lecturer on chemistry in Paris. In 1813, he published the work on which his fame rests, *Traite des poisons* or *Toxicologie generale*. This was the first scientific study of the detection and pathological effects of poisons, and it established toxicology as a distinct forensic field.[20]

Orfila had a rival, François Vincent Raspail, who was retained by the defense in the Lafarge case in 1840. Madame Lafarge was accused of poisoning her husband and indeed had purchased arsenic from an apothecary only days before his death. Chemical tests following the autopsy were inconclusive, however, and Orfila was summoned from Paris. The body was exhumed, and Orfila demonstrated the presence of arsenic in Lafarge's internal organs. In his trial testimony he assured the jury that both his laboratory ware and the cemetery earth were free of arsenic. Raspail, on the other hand, boasted he could extract arsenic from virtually anything—including the judge's chair—using Orfila's procedure. Raspail's testimony was delayed, however, when he fell while riding, and Madame Lafarge was sentenced to the penitentiary. The case is still cited as the first in which an attempt was made to rebut a state's witness by means of an opposing defense expert.[21]

About this time the development of photography also became a technological boon to law enforcement as an aid to the identification of known criminals. As early as 1843 police in Belgium began maintaining files of daguerreotypes for such purposes, and France and the United States followed suit by the 1850s. Still later, the ability to produce multiple prints from negatives greatly facilitated the use of "mug shots" ("mug," the slang term for "face," may have derived from the eighteenth-century custom of fashioning drinking mugs in the form of grotesque human faces).[22]

Even with the assistance of good photographs—a great improvement over memory—mistaken identifications were sometimes made. An example of tragic misidentification is the case of Londoner Adolph Beck who, in 1896 and again in 1904, was mistaken for the swindler William Thomas. Only a postponement and the interim arrest and correct identification of Thomas kept the unfortunate Beck from serving the second prison term. Photographs of the two men do show similarly stout men with walrus mustaches but with otherwise dissimilar facial features.[23] But even photographs did not prevent mistaken identifications. As C.A. Mitchell warned in *The Expert Witness,* published in 1923, "Since a photograph represents only one aspect, and often a false one, of a person at a particular moment, there is a chance of its being mistaken for a portrait of someone else, who in reality is not at all similar, and the possibility of wrong identification may thus be intensified. There have, as a matter of fact, been numerous cases of mistaken identification from photographs."[24]

The first really scientific attempt at identification of criminals was made in 1860 by a Belgian prison warden named Stevens, who began taking measurements of criminals' heads, ears, feet, and the lengths of their bodies. Stevens soon abandoned his imperfect method, but in 1879 Frenchman Alphonse Bertillon (1853-1914) began to develop an elaborate system of anthropometry (the science of measuring the human body). By 1882, his *bertillonage* was being used routinely to identify criminals through tabulation of such factors as height, sitting height, length of outstretched arms, length and breadth of head, length of right ear, and other measurements, as well as photographs. Although this cumbersome and fallible system was replaced after some two decades by fingerprinting, Bertillon's pioneering efforts earned him the sobriquet "the father of criminal identification."[25]

Several individuals contributed to the science of fingerprinting, but credit is given to Francis Galton (1822-1911) for making the first definitive study of the subject. His most important contribution was that he developed a method of classifying fingerprints for filing. His *Finger Prints*, published in 1892, provided the first statistical evidence for the uniqueness of fingerprinting, and it described the fundamental principles that continue to apply to that method of personal identification.[26]

It has been said that "the prototype of today's criminalist was fictional"[27]—Sir Arthur Conan Doyle's Sherlock Holmes. In 1887, when the first Holmes story, *A Study in Scarlet*, appeared, scientific crime detection was still in its infancy, yet Doyle portrayed Holmes not as a mere armchair theorist as was Edgar Alan Poe's Auguste Dupin. Doyle wrote this early account of Holmes's approach:

> He whipped a tape measure and a large round magnifying glass from his pocket. With these two implements he trodded noiselessly about the room, sometimes stopping, occasionally kneeling, and once lying flat upon his face. So engrossed was he with his occupation that he appeared to have forgotten our presence, for he chattered away to himself under his breath the whole time, keeping up a running fire of exclamations, and little cries suggestive of encouragement and of hope. As I watched him I was irresistibly reminded of a pure-blooded, well-trained foxhound as it dashes backwards and forwards through the covert, whining in its eagerness, until it comes across the lost scent. For twenty minutes or more he continued his researches, measuring with the most exact care the distance between marks which were entirely invisible to me, and occasionally applying his tape to the walls in an equally incomprehensible manner. In one place he gathered up very carefully a little pile of grey dust from the floor, and packed it away in an envelope. . . .
>
> "They say that genius is an infinite capacity for taking pains," he remarked with a smile. "It's a very bad definition, but it does apply to detective work."[28]

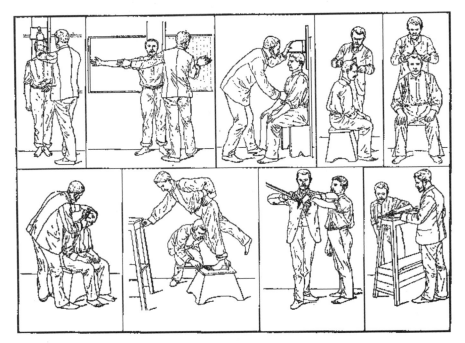

FIGURE 1.1. *Bertillonage*—the science of identification developed by pioneer Alphonse Bertillon—involved recording measurement of various parts of the body. These late-nineteenth-century illustrations show the main aspects of Bertillon's system of anthropometry.

The first real-life "scientific detective" was an Austrian lawyer named Hans Gross (1847-1915).[29] When *A Study in Scarlet* appeared, Gross was making notes for a handbook to be used by his fellow magistrates. The finished work, titled *Handbuch fur Untersuchungsrichter* ("manual for examining magistrates"), appeared in Germany in 1893 and was eventually published in English as *Criminal Investigation*. Insofar as is known, Gross never read the Sherlock Holmes stories, yet his own work seemed to bring the fictional ideas to life. "You had only to open Gross' book to see the dawning of a new age," stated one admiring writer.[30] In addition to advocating forensic medicine, toxicology, serology, ballistics, and anthropometry, Gross had chapters on employment of the mineralogist, ecologist, and botanist. "Dirt on shoes," wrote Gross in a typical sentence, "can often tell us more about where the wearer of those shoes had last been than toilsome inquiries." Gross also coined the term *criminalistics* and later launched the forensic journal *Kriminologie*.[31]

A disciple of Gross (and, admittedly, of Sherlock Holmes) was a Frenchman, Edmond Locard (1877-1966). "I must confess that if in the police laboratory of Lyons we are interested in any unusual way in this problem of dust it is because of having absorbed the ideas formed in Gross and Conan Doyle," he said.[32] This "problem of dust," or trace evidence, was of great interest to Locard, who was educated both in medicine and the law. He set forth the concept known as Locard's Exchange Principle, which states that a cross-transfer of evidence takes place whenever a criminal comes in contact with a victim, an object, or a crime scene. For example, a criminal may leave behind a latent fingerprint and a strand of hair while carrying away from the scene distinctive carpet fibers or other identifiable debris. In the case of three suspected counterfeiters who had been circulating bogus coins, Locard had the suspects' clothing brought to his police laboratory. Careful examination revealed tiny metallic particles in all three sets of clothing, and chemical analyses revealed that the particles had the same metallic elements as the coins. The suspects were arrested and, when confronted with Locard's scientific evidence, confessed.[33]

Given the importance of bloodstains at the scenes of violent crimes, mid-nineteenth century medical-legal experts sought some means of identifying them. Microscopic examination worked on fresh blood, but the distinctive red blood cells broke up and lost their identity as blood dried. Attempts to identify blood chemically were only moderately successful, but the work of two European medical academics, one in Vienna in 1901, the other in Turin in 1915, represented major breakthroughs. Dr. Karl Landsteiner (1868-1943) worked as assistant to a professor of pathology and anatomy at the University of Vienna. After the turn of the century, he conducted experiments in mixing the blood cells and serum of different persons, which led him to the discovery that blood cells can be divided into groups, which were later designated A, B, AB, and O.[34]

Leon Lattes (1887-1954), a professor at the Institute of Forensic Medicine at the University of Turin, developed a forensically useful application of Landsteiner's discovery. On September 7, 1915, a man came to the Institute with two small apparent bloodstains that had mysteriously appeared on the tail of one of his shirts. His wife was berating him over them, and he could not account for them except by speculation. Lattes thought it ironic that a marital dispute rather than a crime had so challenged him, but he set to work to find a means of dissolving the blood and testing it for type. As it turned out, the blood was the man's own, the result of discharge from prostate trouble! Soon, Lattes applied his newfound technique to bloodstains on an accused murderer's coat, exonerating him.[35]

In the field of firearms examination, the U.S. Army colonel Calvin Goddard (1891-1955) laid the groundwork for the individualization of weapons. In the 1920s Goddard refined the process of comparing markings on a bullet taken from a shooting victim with those on one test-fired for forensic examination. Goddard used the comparison microscope for this as well as for comparing the firing-pin marks on shell casings.[36] According to Richard Saferstein, "Goddard's expertise established the comparison microscope as the indispensable tool of the modern firearms examiner."[37]

The greatest handwriting expert of his day, Albert S. Osborn (1858-1936) of New York City, developed the fundamental principles of questioned document examination and so is credited with that field's acceptance by the courts.[38] As one writer observed, Osborn's career "is a history of handwriting testimony in America." Strange as it may seem now, "during his first year as an expert witness he was exposed to the scorn of judges who would not let him set up a blackboard in the courtroom or permit him to introduce, or even mention, evidence revealed by the microscope. Due to the ignorance of early jurors, it was perhaps preferable that they be *told* things instead of being *shown* them, but Osborn changed all that."[39] The same source adds that when Osborn was testifying in the 1930s, the courtroom resembled "a scientific laboratory in which a professor was making a demonstration."[40] Osborn left for posterity his monumental *Questioned Documents,* first published in 1910 and still used as a standard textbook in the field.

Among the important figures in the history of forensic science were those noted for their academic achievement and for the education of others. Dr. Paul Leland Kirk (1902-1970) stands out prominently in this group. Kirk was a student of chemistry and biochemistry and obtained degrees from Ohio State University, the University of Pittsburgh, and the University of California. From 1929 to 1945, at Berkeley, he rose from instructor to professor of biochemistry, taking time off during World War II to conduct plutonium research for the Manhattan Project. By 1934 Kirk began to apply biochemistry to forensic questions, conducting studies by the thousands. He also began to offer courses in criminalistics as part of the curriculum in biochemistry. Eventually, he became the head of the criminalistics department at the University of California's School of Criminology, founded in 1948-50. Kirk's comprehensive work, *Crime Investigation, Physical Evidence, and the Police Laboratory,* quickly became a standard in the field.[41]

Among the more recent superstars of criminalistics are Dr. E. Roland Menzel, professor of physics at Texas Tech University, in Lubbock, Texas, and director of that university's Center for Forensic Studies. Beginning in the 1970s Menzel has pioneered in the application of lasers to criminalistics, especially their use in locating and "visualizing" latent fingerprints and other types of trace evidence, including biological.[42]

Another modern giant is Alec Jeffreys, who, with his colleagues at Leicester University in England in 1985, discovered that portions of certain genes' DNA structure are unique to each person. For that reason Jeffreys named the process used to isolate and read these markers "DNA fingerprinting," now known among criminalists as "DNA typing".[43] Routinely, such biological evidence as blood, semen, tissue, and hair is now being used to exonerate innocent suspects—even those who have been wrongly convicted and sent to prison.

We should, in our historical overview, briefly trace the development of the crime laboratory. The forerunner of such police facilities was the photographic studio, such as the one that originated in Belgium in 1843. Bertillon, the anthropometrist, expanded this facility into an identification bureau that was still far from the suitable laboratory Dr. Edmond Locard dreamed of.

Locard, known for his principle of the cross-transfer of evidence, was disappointed when he visited Bertillon in Paris, and his chagrin continued as he traveled to Lausanne, Rome, Berlin, Brussels, and even to New York and Chicago.[44] Despite a lack of interest, in 1910 Locard was able to use his personal connections and influence to persuade the police prefects in Lyons to provide him with two assistants and makeshift quarters. The latter consisted of two attic rooms above the law courts, the entranceway to which, located on a narrow side street, also led to the prison and the archives. "Every day Locard climbed the steep winding staircase leading to his laboratory four floors up," writes Jurgen Thorwald in his *Crime and Science*. "Twenty years later when he already enjoyed world fame, he still worked up there in somewhat larger quarters, but still under the most primitive conditions, the heating consisting of wretched coal stoves which constantly deposited new layers of soot on the cracked walls."[45] Locard began with only two instruments, an ordinary medical microscope and a small spectroscope (an optical instrument with a prism used to identify substances by the spectrum they emit when burned). These, together with some chemicals and basic lab ware, constituted the world's first scientific crime laboratory, later known as the Lyons Police Laboratory. Within a year Locard had proven the worth of the venture by solving the case of the three coin counterfeiters.

Among the early cases that helped establish Locard's reputation was the murder of a young woman in 1912. Her boyfriend was suspected but seemed to have an airtight alibi; he claimed to have spent the night in question with friends at a distant country house. When Locard was called in he went to the morgue to examine the body and observed strangulation marks on the girl's throat. Next he went to the suspect's cell, where he carefully scraped into small envelopes the debris from beneath the man's fingernails. Back at his laboratory, Locard discovered that the debris consisted of epithelial (skin) cells covered with a pink dust. Under the microscope, the dust proved to be a mixture of rice starch, magnesium sterate, zinc oxide, bismuth, and an iron-oxide pigment known as Venetian red. Locard requested from the police the cosmetics from the victim's room and found that her face powder, specially prepared by a local druggist, had the same ingredients. In the days before mass production of cosmetics, this evidence was enough to induce a confession from the boyfriend, who explained how he advanced the hands of his friend's wall clock to make them think he had been there at the time when the crime was committed.[46]

Following Locard's success, what is now "the oldest forensic laboratory in the United States" was established in 1923 by the Los Angeles Police Department under the direction of August Vollmer.[47] Vollmer was a remarkable man who rose from mailman to chief of police of Berkeley, where it is said he kept a copy of Hans Gross's *Criminal Investigation* on his desk.[48] Vollmer served only a one-year term as chief of the Los Angeles Police Department, but the crime laboratory is an important legacy of his brief tenure. Seven years later, in 1930, the Los Angeles County Sheriff's Department set up its own laboratory, and the following year at Sacramento the California State Crime Laboratory was established. San Francisco and San Diego set up labs in 1932 and 1935, respectively. Each of these was quite small.[49]

Perhaps the first truly significant crime laboratory that could be called a national lab was the Scientific Crime Detection Laboratory, which began at Chicago in 1929 and was soon affiliated with Northwestern University School of Law. The lab originated as a result of the infamous St. Valentine's Day massacre of that same year. When, during the grand jury probe, jurors learned there was no laboratory for testing the numerous bullets and cartridge cases, several influential members raised the funds to establish a permanent forensic laboratory. It was headed by Colonel Calvin Goddard, the pioneer in firearms identification, and appears to have been the first crime laboratory with a firearms examiner on its staff. In 1938 it was transferred to the Chicago Police Department.[50]

The Bureau of Investigation, created in 1908, was reorganized by new director J. Edgar Hoover in 1924, when a national fingerprint file was established by adding fingerprint cards from the federal penitentiary at Leavenworth, Kansas, to existing bureau files. The official United States Crime Laboratory was established by the bureau in Washington, D.C., in 1930, and on November 24, 1932, was equipped to provide forensic science facilities to authorized law enforcement agencies and other government agencies.[51] (The name Federal Bureau of Investigation [FBI] would be adopted in 1935.) Other laboratories were set up in Detroit, Boston, New York, Rochester, New Orleans, and Kansas City. Now numerous cities, counties, and states have crime laboratories or, as in Florida, laboratory systems.[52]

Two years after the FBI laboratory was founded, England's Lord Trenchard established the Forensic Science Laboratory at Hendon. To form a closer contact with the Criminal Investigation Department (CID) it was soon moved to Scotland Yard in London.[53] (The name Scotland Yard derived from the fact that the Metropolitan Police Headquarters, established in 1829 by Sir Robert Peel, in 1842 located its detective force partly at and partly near Great Scotland Yard, a small square where, centuries before, the Scottish ambassadors had lodged. Eventually, the name became attached to the detective force, then to the entire headquarters.[54])

Most forensic laboratories in the United States are publicly—that is, governmentally—operated, usually by law enforcement agencies but occasionally by prosecutors or medical examiners offices, or by departments of public safety. There are also private laboratories, some of which operate as commercial enterprises and others that are affiliated with universities. Since police-operated laboratories are usually unavailable to the defense in court cases, the private laboratories provide an important service in balancing the availability of forensic expertise.[55]

THE MODERN CRIME LABORATORY

The functions of the modern forensic science facility are varied, but, as Paul L. Kirk writes in *Crime Investigation*, "perhaps the most important function of the police laboratory is to train the police investigators as to what constitutes physical evidence and how it is to be found, collected, preserved, and delivered to the proper laboratory investigator." According to Kirk,

As soon as the police investigator discovers how helpful a cooperative criminalist may be to him in increasing his efficiency, any distrust or jealousy of the laboratory worker should cease, and a fruitful and mutually profitable liaison will be established. This will result in more effective police work, which will benefit the entire force and the political subdivision it serves. Public relations will improve, police practice will be increased, and an atmosphere will be created in which confidence and respect, as well as more immediate personal advantages, will accrue to the force. The laboratory investigator and the police officer must always keep in mind that they are not competitive but complementary in their functions. The laboratory cannot produce unless the officer makes it possible, and the officer can solve many more crimes if he utilizes the laboratory to the fullest extent. . . .

The study of physical evidence has a twofold purpose. *First,* and most important, *it is often the decisive factor in determining guilt or innocence.* Thus, the testimony of the scientific expert may be sufficient to determine the final decision of a court. It can do this by supplying the demonstrable facts, thus resolving discrepancies in ordinary testimony, and amplifying the information of the court to a point at which a true and just decision of guilt or innocence may be rendered, unclouded by divergent statements of uncertain or perhaps prejudiced witnesses. *Second, the study of physical evidence can be a material aid in locating the perpetrator of a crime.*

Physical evidence is often very useful to the police investigator before he has a suspect in custody, or in fact, before he even has suspicions of a possible perpetrator. If, for instance, the laboratory can describe the clothes worn by the criminal, give an idea of his stature, age, hair color, or similar information, the officer's search is correspondingly narrowed.[56]

How the crime laboratory is set up depends on several factors, including the social nature and the size of the community that it serves, the anticipated case load, and the available facilities and funding. Cunliffe and Piazza recommend the following general divisions for crime laboratories. (Of course, larger laboratories would have more extensive capabilities.)[57]

Photography Section. This will include the necessary equipment for crime scene photography, in addition to a studio for identification photos and other in-house photographic work, and a darkroom for processing all of the laboratory's film and prints.

Evidence Storage Section. This will consist of a secure area for evidence storage, and there will be employed a suitable receiving procedure to maintain the accountability of every evidential item and to ensure the continuity of the *chain of custody*. Every person who handles evidence must be accounted for to prevent questions about the integrity and even the authenticity of evidence.[58]

Identification Section. This will include equipment and facilities for recording fingerprints of persons, both inside and outside the lab; for locating and recording latent fingerprints at crime scenes and on items of evidence transported to the lab; and for classifying and filing standard fingerprint cards. This section may also be charged with making casts of shoe and tire impressions.

Chemistry Section. Here a variety of tests and examinations will be conducted including drug identification, toxicology analysis, examination of body fluids such as blood and semen, and the restoration of serial numbers.[59] Larger laboratories may have separate sections for drugs, toxicology, and serology.

General Examination Section. Because some areas overlap, this will be closely associated with the chemistry section. Here microscopical examinations are performed on soil, glass, paint, metal, explosive residues, hairs and fibers, and other trace evidence. Other types of examinations of physical evidence such as matching pieces of broken glass may also be conducted here. Ideally, there will be a separate section for document examinations.

Firearms Section. In addition to providing the capability of identifying firearms, bullets, and shell cases, and of associating spent bullets and cases with the weapons from which they were fired, this section should have a separate shooting room for firing test bullets and comparing gunpowder and shot patterns in making distance determinations. As necessary, this section should coordinate work with the chemistry, photography, and identification sections.

Instrument Section. Depending on need and resources, this section may be extensive or somewhat limited. Instruments ideal for the well-equipped laboratory include a gas chromatograph; an x-ray diffraction unit; an emission spectrograph; a mass spectrometer; infrared, ultraviolet, and atomic absorption spectrophotometers; equipment for thin-layer chromatography and electrophoresis; and a sound spectrograph.

Crime Scene Search Section. Today, large city and state crime laboratories often have sufficient caseloads and revenues to be able to develop teams of criminalists or specially trained technical investigators who search crime scenes and collect evidence for processing at the laboratory. That is the mission of this laboratory section.

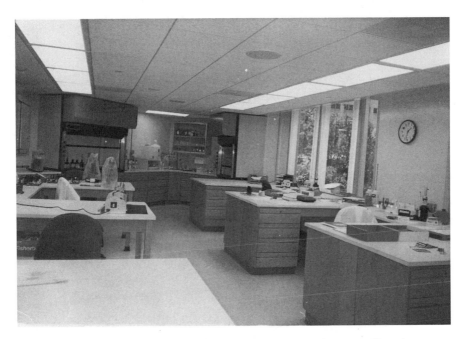

FIGURE 1.2. A typical work area of a crime laboratory's Drug Chemistry Section, illustrating workbenches and frequently used instruments. (Courtesy of Chris Watts.)

Although there is the tendency today for criminalists to specialize, Cunliffe and Piazza caution that "it is a mistake for a laboratory to become over-specialized. Since almost everything can become physical evidence at one time or another, scientific investigation can encompass many different facets of science, including elements of physics, chemistry, biology, geology, and metallurgy." For this reason, they conclude that "it is desirable for the criminalist to be something of a scientific generalist."[60]

Whenever specialized services are needed but are lacking at the local laboratory, the FBI laboratory can assist if a rather long wait is acceptable. Effective July 1, 1994, however, the FBI ceased to provide scientific examination in cases of property crime such as burglary, auto theft, non-fatal traffic accidents, fraud, or theft under $100,000, except cases that involve or were intended to cause personal injury.[61]

One of the greatest resources of the FBI is their extensive reference collections used to help solve crimes. For example, for crimes involving firearms, they have available a Firearms Reference Collection containing more than two thousand handguns and over eight hundred shoulder weapons, used for the identification of gun parts and the locating of serial numbers; a Standard Ammunition File with over ten thousand samples of foreign and domestic specimens of ammunition; and a Reference Fired Specimen File consisting of test bullets and cartridge cases fired from weapons submitted to the laboratory. For questioned document examination they have on hand a Typewriter Standards File containing original specimens of typewriting from numerous and domestic machines; a Watermark Standards File, which indexes watermarks and brand names used by paper manufacturers and helps trace the origins of paper; a Safety Paper Standards File that helps determine the manufacturers of "safety" paper used for checks; the Checkwriter Standards File, which contains original impressions of checkwriters so they can be identified as to make and model; and an Office Copier Standards File, which aids in determining the manufacturers of photocopiers and duplicators.

The FBI also maintains reference files of inks; a special Anonymous Letter File with the handwriting, handprinting, and typewriting of extortionists, confidence men, and kidnappers; the National Fraudulent-Check File, which consists of photographic copies of the works of "bad-check artists"; and a Bank Robbery Note File.

Other fields of forensic science are well represented in the FBI laboratory. Investigations may be aided by a vast Tire Tread File of patterns furnished by manufacturers; a Shoe Print File; a National Vehicle Identification Number (VIN) Standard File, which allows lab personnel to determine the authenticity of a VIN number on a stolen vehicle; the National Automobile Altered Numbers File, which includes specimens of fake VINs found during investigations, the National Motor Vehicles Certificate of Title File with authentic samples from each manufacturer and state issuer as well as photographic copies of fraudulent titles and stickers; and the Explosive Reference Files with technical data and known standards of various explosive items and bomb components. There is even a Cigarette Identification File, which is used to identify cigarette butts found at crime scenes, and a Pornographic Materials File. In addition, the FBI laboratory maintains the Automobile Paint File, which can be important in hit-and-run cases; an extensive Hairs and Fibers Collection; Safe Insulation Files; Blood Serum Files; and an Invisible Laundry Mark File.[62]

The U.S. Treasury Department also maintains reference files, notably, since 1968, a complete "library" of every available commercial pen ink. Inks are cataloged according to dye patterns developed by the technique known as thin-layer chromatography. In one case, for example, it was possible to demonstrate that a document dated 1958 was actually back-dated, since a dye in the ink had not been synthesized until a year later.[63]

Two important regulation processes have now been implemented within the forensic sciences: certification and accreditation. Several boards, including the American Board of Criminalists, certify individual forensic scientists as to their level of knowledge, skills, and expertise in specified areas. A critical part of the certification process is proficiency testing of applicants. Similarly, strenuous efforts are being made by the American Society of Crime Laboratory Directors (ASCLAD) to improve the quality of crime laboratories, including local, state, and federal facilities, by accreditation. One of ASCLAD's successes is that all federal laboratories are now obligated to seek the board's accreditation.

In closing this introduction to the crime laboratory in particular, and of criminalistics in general, it is well to consider these words from an article in the *British Journal for the Philosophy of Science:* "The scientist is indistinguishable from the common man in his sense of evidence, except that the scientist is more careful. The increased care is not a revision of evidentiary standards, but only the more patient and systematic collection and use of what anyone would deem to be evidence."[64]

NOTES

1. This discussion is abridged from Joe Nickell, "Investigatory Tactics and Techniques," in *Missing Pieces,* ed. Robert A. Baker and Joe Nickell (Buffalo: Prometheus, 1992), 69-78.

2. Harold Tuthill, *Individualization Principles and Procedures in Criminalistics* (Salem, Oregon: Lightning Powder, 1994), 28-30.

3. Peter R. De Forest, R.E. Gaensslen, and Henry C. Lee, *Forensic Science: An Introduction to Criminalistics* (New York: McGraw Hill, 1983), 4.

4. Quoted in Tuthill, *Individualization,* 6.

5. Quoted in ibid., 10.

6. Ibid., 17-20.

7. Ibid., 21-22.

8. Ibid., 23.

9. "Identification of Typewriting," forensic instruction manual for course in Scientific Crime Detection (Chicago: Institute of Applied Science, 1962), lesson 2, 18-21.

10. Stuart Kind and Michael Overman, *Science Against Crime* (Garden City, N.Y.: Doubleday, 1972), p. 27.

11. De Forest, Gaensslen, and Lee, *Forensic Science,* 32-38.

12. Ibid., 47.

13. Ibid.

14. Thomas L. Bohan and Erik J. Heels, "The Case Against *Daubert:* The New Scientific Evidence 'Standard' and the Standards of the Several States," *Journal of Forensic Sciences,* vol. 40, no. 6 (Nov. 1995): 1030-44. (The quotation is from the abstract.)

15. Quoted in Richard Saferstein, *Criminalistics: An Introduction to Forensic Science,* 5th ed. (Englewood Cliffs, N.J.: Prentice Hall, 1995), 16.

16. Richard Saferstein, "Forensic Science: Winds of Change," in *Chemistry and Crime: From Sherlock Holmes to Today's Courtroom,* ed. Samuel M. Gerber (Washington, D.C.: American Chemical Society, 1983) 39.

17. Kind and Overman, *Science Against Crime,* 12.

18. *Encyclopedia Britannica,* s.v. "Medical Jurisprudence."

19. Saferstein, *Criminalistics,* 4.

20. Ibid.; *Encyclopedia Britannica,* s.v. "Mathieu Joseph Bonaventure Orfila." Saferstein gives the date of Orfila's treatise as 1814.

21. De Forest, Gaensslen, and Lee, *Forensic Science,* 9.

22. Joe Nickell, *Camera Clues: A Handbook for Photographic Investigation* (Lexington: Univ. Press of Kentucky, 1994), 70.

23. Ibid., 72.

24. C.A. Mitchell, *The Expert Witness* (New York: D. Appleton, 1923), 24.

25. Nickell, *Camera Clues,* 69; Saferstein, *Criminalistics,* 4.

26. Saferstein, *Criminalistics,* 4.

27. De Forest, Gaensslen, and Lee, *Forensic Science,* 12.

28. Arthur Conan Doyle, *A Study in Scarlet* (1887; reprinted in *The Complete Sherlock Holmes,* Garden City, N.Y.: Garden City Books, 1930), 22-23).

29. De Forest, Gaensslen, and Lee, *Forensic Science,* 12.

30. Jurgen Thorwald, *Crime and Science: The New Frontier in Criminology* (New York: Harcourt, Brace & World, 1966), 234-35.

31. Ibid.; Saferstein, *Criminalistics,* 5.

32. Quoted in Henry Morton Robinson, *Science Catches the Criminal* (New York: Blue Ribbon, 1935), 201.

33. Saferstein, *Criminalistics,* 5-6.

34. Ibid.; Thorwald, *Crime and Science,* 32-35.

35. Thorwald, *Crime and Science,* 38-42.

36. Ibid., 444.

37. Saferstein, *Criminalistics,* 5.

38. Ibid.; Robinson, *Science Catches the Criminal,* 119-20.

39. Ibid.

40. Ibid., 120.

41. Ibid., 14; Thorwald, *Crime and Science,* 149-0.

42. Keith M. Beesley, Savvas Damaskinos, and A.E. Dixon, "Fingerprint Imaging with a Confocal Scanning Laser Macroscope," *Journal of Forensic Science,* vol. 40, no. 1, (Jan. 1995): 17; Barry A.J. Fisher, *Techniques of Crime Scene Investigation,* 5th ed. (New York; Elsevier, 1992), 111.

43. Saferstein, *Criminalistics,* 383.

44. Thorwald, *Crime and Science,* 282.

45. Ibid., 283.

46. Ibid., 284-85, 286.

47. Saferstein, *Criminalistics,* 6.

48. Thorwald, *Crime and Science,* 149; Colin Wilson, *Clues: A History of Forensic Detection* (1989; reprint, New York: Warner, 1991), 461.

49. De Forest, Gaensslen, and Lee, *Forensic Science,* 13.

50. Ibid., 14-15.

51. *Encyclopedia Britannica,* s.v. "Federal Bureau of Investigation."

52. De Forest, Gaensslen, and Lee, *Forensic Science,* 15.

53. Sir Harold Scott, *Scotland Yard* (New York: Random, 1955), 140.

54. John Wilkes, *The London Police in the Nineteenth Century* (Minneapolis: Learner, 1977), 28.

55. De Forest, Gaensslen, and Lee, *Forensic Science,* 15.

56. Paul L. Kirk, *Crime Investigation,* 2d ed., ed. John I. Thornton (New York: John Wiley, 1974), 3-4.

57. Frederick Cunliffe and Peter B. Piazza, *Criminalistics and Scientific Investigation* (Englewood Cliffs, N.J.: Prentice-Hall, 1980), 7-9.

58. Saferstein, *Criminalistics,* 43.

59. Ibid., 456.

60. Cunliffe and Piazza, *Criminalistics and Scientific Investigation,* 5.

61. "Notice to State and Local Crime Laboratories," FBI, 1994.

62. Charles R. Swanson Jr., Neil C. Chamelin, and Leonard Territo, *Criminal Investigation,* 4th ed. (New York: McGraw-Hill, 1988), 239-40, (See also C.B. Colby, *F.B.I.: The 'G-Men's' Weapons and Tactics for Combating Crime* (New York: Coward-McCann, 1954), 16-30.

63. Saferstein, *Criminalistics,* 486.

64. Quoted in Tuthill, *Individualization,* 8.

RECOMMENDED READING

Kind, Stuart, and Michael Overman. *Science Against Crime.* Garden City, N.Y.: Doubleday, 1972. A popular introduction to the realm of forensic science, well illustrated.

Paul, Philip. *Murder Under the Microscope: The Story of Scotland Yard's Forensic Science Laboratory.* London: Futura, 1990. Popular case study of forensic methods in historical context.

Robinson, Henry Morton. *Science Catches the Criminal.* New York: Blue Ribbon, 1935. A dated but nevertheless useful survey of the forensic sciences, with much historical information on pioneers such as Locard, Goddard, and Osborn.

Saferstein, Richard. *Criminalistics: An Introduction to Forensic Science,* 5th ed. Englewood Cliffs, N.J.: Prentice Hall, 1995. A serious, updated, comprehensive look at the work of the criminalist, including the latest information on DNA, arson and explosives, and voice examination.

Thorwald, Jurgen. *Crime and Science: The New Frontier in Criminology.* New York: Harcourt, Brace & World, 1966. An excellent history of forensic serology (part 1) and forensic chemistry and biology (part 2) that narrates the progress of these sciences through the major cases that challenged them.

Tuthill, Harold. *Individualization Principles and Procedures in Criminalistics.* Salem, Oregon: Lightning Powder, 1994. Introduces the field of criminalistics and clarifies the crucial terms *identify* and *individualize,* focusing on the latter as the preferred term for demonstrating the uniqueness of items of evidence.

Wilkes, John. *The London Police in the Nineteenth Century.* 1977. Reprint, Minneapolis: Learner, 1984. Brief, readable and well-illustrated discussion of the development of the London police force; a useful background to the role of science in crime detection.

2

CRIME-SCENE INVESTIGATION

In crimes of violence and in burglary, the scene of the crime may be the most important aspect of the investigation. When Charles E. O'Hara states in his *Fundamentals of Criminal Investigation* that "there is not only the effect of the criminal on the scene to be considered, but also the manner in which the scene may have imparted traces to the criminal,"[1] he is restating Locard's Exchange Principle. Skilled, painstaking work is required to make this principle effective in solving crimes. Crime-scene investigation consists of certain *preliminaries,* followed by *documentation,* then the *collection and preservation* of the evidence. Only then may crime reconstruction be possible. Finally, certain *legal considerations* must be followed.

PRELIMINARIES

Nowhere is the adage "First things first" more appropriate than at the scene of a crime. Both the police officer, who is likely to be the first official to arrive at the scene, and the investigator, who typically arrives soon thereafter, must see that vital information is not lost. The following preliminary actions are recommended.[2]

1. Deal with any emergency situation that is found (such as a person who requires first aid or a weapon that needs to be secured) regardless of the following guidelines.

2. Identify the person who reported the crime or disturbance to the police and if possible detain the person for questioning. (The individual may have valuable information to impart and may even become a suspect.)

3. Attempt to determine the perpetrator of the crime if that can be immediately determined by inquiry or direct observation.

4. Detain everyone who is present at the scene, including eyewitnesses to events leading up to the crime, to the crime itself, or to its aftermath or people in possession of other information that must be garnered while fresh.

5. Summon whatever assistance may be necessary. Police officer(s) may need additional personnel to assist with crowd control; investigator(s) may require assistance in safeguarding the area.

6. Secure the scene by the issuance of necessary orders and by physical isolation of the area. Unauthorized persons, such as spectators, newspaper reporters, and television crews, must be kept well away from the premises; people milling about a crime scene may obliterate crucial evidence. Uniformed officers can be used to rope off the scene and to post "Police Line Do Not Cross" signs or "Stop: Crime Scene Search Area" cards where appropriate, or specially imprinted tape may be used to cordon off the area. Police officers should be instructed that all witnesses they encounter should immediately be referred to an investigator.

7. Separate the witnesses so that their statements are not influenced by one another.

8. Secure and investigate any additional crime scene, such as a place from which a body was moved or where a vehicle connected to the crime is discovered.[3]

9. If it is known there is no emergency, avoid rushing to the area of focus (for example, the body in a homicide case); minute but valuable evidence may thereby be altered or even destroyed. (Of course, in the process of rendering first aid, checking for signs of life, or handling some other emergency, a police officer or emergency medical technician may alter the scene and inadvertently destroy evidence, but that is simply the unavoidable consequence of a more important duty and no blame is attached.)[4]

10. Refrain from moving, or even touching, any object at the scene. It is impossible to restore an item of evidence to its original position once it has been moved.

11. If assistants are available, assign each one definite duties so as to minimize confusion and eliminate duplication of effort.

FIGURE 2.1. Array of equipment typically found in a crime-scene vehicle. (Courtesy of Ed Hobson, Orlando Police Department, Orlando, Florida.)

Ideally, a crime-scene team such as the following should be assigned to conduct the search:

The *officer in charge* assumes overall responsibility for the effectiveness of the work, gives directions, and delegates assignments.

An *assistant* carries out the directions of the officer in charge.

The *photographer* takes the necessary photos of the scene as well as of each individual item of evidence as it is discovered.

The *sketcher* prepares a rough draft and later a finished drawing of the crime scene.

The *master note taker* records in shorthand such observations and descriptions as others provide, noting the time at which a discovery is made and by whom, and keeps a well-ordered log of the activities.

The *evidence person* collects, tags, and preserves each item of evidence.

The *measurer* takes general measurements of the scene, such as will be needed for the crime-scene sketch, and also locates by a suitable means such as a coordinate system every significant object and each item of evidence that is present.

The *section leader* takes charge of personnel in a section (usually made up of six individuals) whenever a large scene must be searched in teams and makes arrangements for collecting, preserving, and transporting any evidence that is discovered[5].

FIGURE 2.2. To prevent contamination of the crime scene, specially marked yellow tape is frequently used to provide a barrier to unauthorized persons. (Courtesy of Don Ostermeyer, J.L. Bunker & Associates, Ocoee, Florida.)

As noted, this team represents the ideal. In reality, the majority of crime scenes are usually handled by one or two technicians and possibly a supervisor.

DOCUMENTATION

The documentation of a crime scene consists of photographs, the crime-scene sketch, and notes.

Photographs. A good photographic record helps document the facts and physical circumstances at the crime scene, records evidence that because of size or other reasons cannot easily be brought to court, permits reconstruction of the crime, and reveals evidence that might otherwise be missed. It can also assist in refreshing the investigator's mind at any time, especially in preparation for court testimony.[6]

In order to be as effective as possible, crime-scene photographs should meet certain fundamental criteria. First, since time is of the essence, photographs should be made before other stages of the investigation are carried out. Before anything is touched or removed, photographs should be taken of the entire area, with views from every possible angle, together with close-ups of anything that might have a bearing on solving the crime. To accommodate this need to make photographs immediately, the necessary equipment must always be kept ready.[7]

Proper views are especially important. The camera position used for each exposure should be recorded on the crime-scene sketch or in the report made from on-site notes. To be as comprehensive and effective as possible, photographs should be taken in *overlapping* segments that flow in one direction around the room or other area, with the camera approximately at eye level (unless a tripod is employed). People should not appear in the photographs. Above all, a correct photographic perspective must be maintained—that is, care must be taken to prevent distortions from reducing or even destroying a photograph's evidentiary value.[8]

Some courts object to any extraneous items—such as a paper arrow or even a ruler for scale—appearing in crime-scene photographs. Such articles are thought to constitute a "modification" of the scene. Because of this attitude, most authorities advise taking pictures both with and without such aids. For this and other reasons, it is wise to make more photographs than may seem strictly necessary; the cost is relatively small, and making additional photos once the scene is disturbed by the collection of evidence is useless.[9]

Before they are removed from the crime scene, items of evidence, such as an apparent murder weapon or a blood stain, should be photographed first at close range (to show the object clearly), then at a distance sufficient to place the object in proper context. Some items of evidence require special photographic procedures that will be discussed in later chapters. According to the textbook *Criminal Investigation,* the investigation of a homicide committed in a house would require the following photographs:[10]

1. The line of approach to and flight from the scene

2. Significant adjacent areas, such as the yard of the house in which the homicide occurred

3. Close-up photographs of the entrance and exit to the house used by the suspect or those most likely to have been used if these are not obvious

4. A general scenario photograph showing the location of the body and its position in relation to the room in which it is found

5. At least two photographs of the body, at 90-degree angles to each other, with the camera positioned as high as possible, pointing downward toward the body

6. As many close-ups of the body as needed to show wounds or injuries, weapons lying nearby, and the immediate surroundings

7. The area underneath the body and under each item of evidence immediately after its removal, if there is any mark, stain, or other apparent alteration produced by the presence of the body or evidence

8. All blood stains, preferably in color

9. All latent fingerprints after dusting and before they are lifted and the weapons on which the prints were found, showing each print's relationship to the general surroundings. (Latent fingerprints likely to be destroyed by lifting always must be photographed, even when it is not standard practice to shoot all fingerprints before handling.)

In the case of buried bodies, any evidentiary item discovered en route to the site—such as tire tracks or articles of clothing—should be photographed *in situ* (in their original place). After photographing the site, investigators should use stakes and string to superimpose on the site a grid, which is usually oriented to north. The scale drawing is prepared with corresponding squares; in addition to this plane (top) view, an elevation (side) view is also prepared so that items can be located as to depth. Soil should be removed carefully in even layers of about four to six inches, using a hand-trowel or flat-bladed spade. As dirt is removed it should be sifted, first through a quarter-inch mesh screen followed by sifting through ordinary window screening. Items should be photographed *in situ* as discovered—both with ruler and north indicator and without. Items found on the screens should be photographed there (with scale, of course, but not with north indicator); they should never be returned to the excavation for photographing; that would constitute false documentation.[11]

The Crime-Scene Sketch. The sketch complements the photographs and notes made during the search of the crime scene. It has several advantages: it can display areas that photographs cannot, such as the floor plan of a house; it ties together multiple, rather fragmented, views and eliminates distortion caused by perspective; it eliminates unnecessary detail so that essential elements are shown clearly; it illustrates the location of evidence, such as a latent fingerprint, that is not visible in a photograph; and it can be used to plot accurately measurements made at the scene.[12]

FIGURE 2.3. Reflected ultraviolet photography can be carried out in the crime laboratory as well as at the crime scene. Specialized lenses, filters, and light sources are needed to "visualize" (as criminalists say, meaning to render visual) what is frequently invisible to the eye.

The rough sketch is just that; it need not be a work of art. However, neat rendering with a straight edge is desirable. Graph paper is an aid to accuracy and proportion and can be easily affixed to a clipboard. Little else is needed, beyond a steel tape measure and a magnetic compass (for indicating directions), all of which will fit into the sketcher's briefcase. Use of a soft lead pencil and eraser will make corrections easier; in *Fundamentals of Criminal Investigation,* however, Charles E. O'Hara cautions that "no changes should be made on the original sketch after the investigator has left the scene."[13]

To locate points on a sketch, some effective method must be used, such as that of *rectangular coordinates*. For this, two perpendicular lines are used—say, the walls in a room. The original position of an item of evidence, such as a pistol, can then be located by measuring at right angles its distance to each wall. Two intersecting streets may similarly be used for exterior work. Or a system *of polar coordinates* may be used. A fixed point is chosen—for example, an electric pole—and a desired object is located by distance and compass reading; positions can later be plotted accurately on the finished drawing by use of a protractor.[14]

FIGURE 2.4. Luminescent fingerprint powders, in conjunction with forensic light sources, are used to develop shoe impressions as well as latent fingerprints. The luminescent shoe impression shown here was developed on vinyl flooring. Note the "L"-shaped scale, which will be used to enlarge the photograph to "life size" (i.e., the actual size of the imprint) for subsequent examination.

An ordinary sketch may be purely two-dimensional. Should it be necessary to represent three dimensions—for example, to locate something on a wall—a cross-projection sketch is utilized. This sketch simply extends the wall from the floor plan as if the wall has been flattened. If necessary, all four walls can thus be shown, arranged around the floor plan to form a cross.[15]

A finished drawing is prepared from the on-scene sketch and notes. The drawing is primarily intended for presentation in court; therefore, good appearance can be a virtue, and the drawing may be prepared by an expert draftsman, under the direction of the investigator. A good final drawing will include the following information:

1. The name and, if applicable, the rank and shield number of the investigator who prepared it

2. The date and time of the sketch, the crime classification, and case number

3. The full name of anyone who assisted in taking measurements

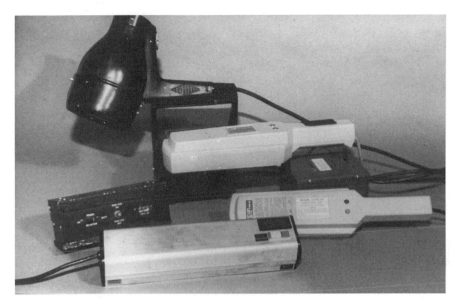

FIGURE 2.5. Forensic light sources are frequently used at a crime scene to detect a variety of physical evidence, such as latent fingerprints or shoe impressions, as well as other trace evidence. Shown here are several types of ultraviolet lights that are easily transported to the location.

4. The crime scene's address, location inside a building, nearby landmarks, and an arrow indicating north

5. The drawing's scale (for example, "one inch equals one foot"; if the drawing is not to scale there should be an indication, such as "not to scale; dimensions and distances tape measured")

6. The location—with accurate measurements—of every major item of evidence and all other important features at the scene

7. A legend of the symbols used to identify points of interest on the drawing.

Color may be used but not to the point of confusion.[16] Accuracy is of the utmost importance. An erroneous measurement creates doubts about the accuracy of other work.

Notes. As indicated earlier, notes for a crime scene are ideally maintained by a designated member of the investigative team. This individual will keep a log that records the time of discovery of the crime, the identity of the person reporting it, and observations and descriptions provided by others. Notes should, for example, include the date and time photographs

FIGURE 2.6. Bloodstain evidence must be photographed to illustrate the "natural" condition of the evidence prior to taking samples or using enhancement techniques. This photograph of a bloody shoe impression was taken prior to reflected ultraviolet photography (see next illustration).

are taken, the name of the person who discovers each piece of physical evidence and its description and location, the name of the person packaging and marking the evidence, and the subsequent disposition of each marked item.[17] Richard Saferstein sounds a note of caution in *Criminalistics*: "The note taker has to keep in mind that this written record may be the only source of information for refreshing one's memory months, perhaps years, after a crime has been processed. The notes must be sufficiently detailed to anticipate this need."[18] Actually, this need for detail as a memory refresher applies to the entire documentation process. Saferstein also recommends audio and video tape; "detailed notes can be taped much faster than they can be written" and "narrating a videotape of the crime scene . . . has the advantage of combining note taking with photography." Saferstien notes, however, that "at some point the tape *must* be transcribed into a written document."[19]

COLLECTION AND PRESERVATION

A thorough examination of the crime scene is essential, so a plan of search should be devised that will insure complete coverage of the area.

FIGURE 2.7. This reflected ultraviolet photograph of the bloody shoe impression shown in the previous illustration reveals additional shoe impressions and artifacts not previously seen.

Indoor scenes will quite naturally be searched on a room-by-room basis, with each room subdivided as necessary into squares or sectors and, where desirable, each sector further subdivided. This is called the *zone method.*

A related technique, the *grid method,* is the standard procedure for covering large areas. It is implemented in a number of variations, a common approach being for searchers to walk in parallel rows, either horizontally or vertically, shifting rows to new locations as necessary until the search area has been completely covered; usually the personnel then repeat the search, this time in rows that traverse at right angles to the first.[20] The searchers are, of course, looking for anything that might provide a clue in the case at hand; more specifically, they are seeking anything that would shed light on the *corpus delicti,* or the fact that the crime in question was indeed committed (for example, an empty money bag at the rear of a burgled store); the perpetrator's *modus operandi* or "M.O." ("method of operation," such as use of a glass cutter to effect an illegal entry);[21] and the perpetrator's identity (for instance, an ejected shell case or anything else that may be traced to an item in a suspect's possession).[22]

In collecting evidence, care should be taken to acquire adequate samples sufficient for a battery of tests by the laboratory and additional

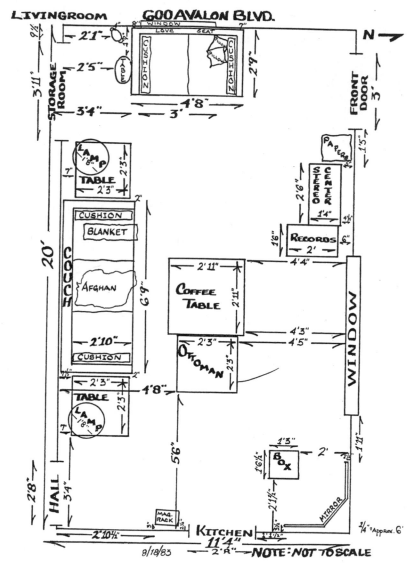

FIGURE 2.8. Diagrams of a crime scene supplement investigator's notes and illustrate the relative layout of the area. (Courtesy of Don Ostermeyer, J.L. Bunker & Associates, Ocoee, Florida.)

analyses by the defense, with a quantity left over for any future review or retrial. In collecting blood and similar stains and traces, the criminalist should also collect *control samples*—that is, specimens of any foreign substance or background material that might have contaminated the stain. Such controls can be tested independently of the stain to determine if any contamination exists.

The integrity of samples must also be maintained. It is imperative that any evidence sample be kept from contact with any other sample or contaminating material. O'Hara writes:

> For example, if a tool impression is found on the door of the house during an investigation of a burglary and if a jimmy is found in association with a suspect, there is a temptation for the investigator to experimentally determine whether the blade of the tool fits the impression by placing the tool against the door. The result is a contamination of any paint traces that may have lain on the blade of the tool and which would have served as stronger evidence of its use in the commission of the crime than would the impression alone. A less conscionable error is the placing of the two poorly wrapped samples of known and unknown in the same envelope. To maintain individuality each piece of evidence should be separately wrapped and should not share the same container unless all danger of mingling is removed by the employment of strong protective coverings or partitions.[23]

To preserve evidence samples, they should be kept from high temperatures and should be placed in storage as soon as possible. Some types of perishable materials require preservatives, but these should be used only with expert knowledge. Evidence must be carefully protected when being prepared for transfer to the laboratory. Trace evidence—hairs and fibers, or dirt, filings, or other particulate matter—should be picked up with a forceps or spatula and placed in bottles with snap-on tops or loosely wrapped in filter paper and placed in a pillbox. Documents and fingerprints on paper should be placed, without folding, in cellophane envelopes. Bullets and shell cases should be surrounded with cotton in separate pill bottles or boxes. Clothing should have any stains or other points of interest widely encircled with chalk to call the attention of the laboratory expert to them. Garments should be folded carefully with clean white paper inserted between stains, and placed in a suitable clean box for transportation. Liquid blood should be deposited in a test tube using an eye dropper, and saline solution should be added in a ratio of about one part saline to five parts blood. Firearms, knives, and tools may be secured with string to a board having perforations for that purpose; if fingerprints have already been lifted from such items, they may be wrapped in paper and placed in a sturdy box. In removing paint, a wooden implement such as a tongue depressor should be used since a metal tool could leave traces that would interfere with spectrographic analysis. If necessary, a steel scalpel may be used and then delivered with the samples to the laboratory expert.[24]

All evidence[25] must be marked for identification and logged. Bullets, shell cases, and any objects of a cubic inch or larger should generally be inscribed with the initials of the investigator who finds or receives the evidence, with the marking placed away from any potential evidentiary traces. A sharp stylus for scratching and a pen for writing are the customary means of marking evidence. Marked or unmarked items are further protected by being sealed in containers in such a way that the seals must be broken for the containers to be opened. The investigator's name or initials and the abbreviated date should be written in ink on the seal. An adhesive label should then be affixed to the box or envelope, or a tag attached to a marked rifle, bearing the case information:

1. Case number
2. Date and time the item was found
3. Item's name and description
4. Location of the item when it was discovered
5. Signature of the investigator who discovered the item
6. Names of witnesses to the discovery[26]

Special care must be taken in transporting the evidence so that it is not altered or damaged in transit. Finally, the evidence should be deposited in a secure evidence storage facility, and it should remain there except when it is removed for some legitimate purpose such as laboratory examination or for use as evidence at a trial. Each time an item is removed or deposited, the following should be recorded in ink in a log:

1. Date evidence was received in storage
2. Case file number
3. Case title
4. Person from whom or place from which the item was received
5. Name of the staff person who received the evidence
6. Complete description of the evidence including such identifying data as serial number
7. Disposition, whether to an individual (named) or other
8. Signature of the officer in charge of the evidence room[27]

FIGURE 2.9. A crime-scene technician uses a metal detector to search for evidence at an outdoor site. (Courtesy of Ed Hobson, Orlando Police Department, Orlando, Florida.)

Attention to these details will protect evidence to help ensure that justice is meted out.

CRIME RECONSTRUCTION

Following documentation of the crime scene and the collection of physical evidence, the investigator should attempt to determine the sequence of events that constituted the crime. This attempt to ascertain the circumstances is known as *reconstruction*. It may range from a mental exercise to a videotaped recreation of theorized events.

The first step is generally to reconstruct the scene's physical appearance based on the indications of evidence, both physical and eyewitness. Whenever conditions of weather and/or lighting are pertinent, the reconstruction should be carried out under similar circumstances. Witnesses should be directed to resume their positions and then to reenact in proper sequence their movements at the time. Other persons may take the positions and play the roles of the participants.[28]

After the physical reconstruction, the investigator attempts to reach conclusions about various, perhaps conflicting, testimony and physical evidence. For example, in the case study at the end of this chapter, the statements Dr. Jeffrey MacDonald made concerning the violent murders of his wife and children by assailants with whom he allegedly struggled were at variance with physical evidence documented and collected at the crime scene.

In *Fundamentals of Criminal Investigation,* O'Hara states that "in reconstructing the actions of the criminal, the investigator should test his theory for logic and consistency. A theory should not be rejected merely because the investigator might not under the circumstances behave in a similar manner. The study should be conducted from the point of view of the mentality of the criminal. No assumptions should be made concerning actions which are not supported by evidence. The theory finally developed by the investigator should provide a line of investigative action but should not be pursued in the face of newly discovered facts which are not consistent with it.[29]

Blood Patterns. The nature of blood evidence plays a crucial role in many reconstructions of violent crimes, including the MacDonald case. The different ways bloodstains appear when the blood leaves the body and is transferred to a surface has led to a criminalistics subspecialty known as blood-pattern analysis. Blood-pattern evidence may reveal, in addition to whose blood is present, the type and order of injuries, the type of weapon that caused each injury, and whether the victim was in motion when an injury occurred or whether he or she moved—or was moved—afterward.[30]

Blood is an aerodynamically uniform material whose behavior is not significantly affected by temperature, atmospheric pressure, or humidity. In blood drops, scalloped edges generally indicate a greater distance of falling than do perfectly round edges, given the same target surface. Distance estimates, however, are of no value unless the effect of the surface upon which the blood falls is well known. The directionality of a small bloodstain or droplet is easily determinable from its uniform teardrop shape, with its tail pointing in the direction of travel. Confusion may result, however, from what are called cast-off droplets—secondary spatter from a larger drop; these droplets have longer tadpole shapes with the sharper ends pointing toward the direction from which they came. If no larger drop is present among hundreds of drops less than one-eighth of an inch across, one may conclude that the pattern resulted from an impact.[31]

Judith Bunker is a nationally known blood-pattern analyst who has assisted the present authors in various cases. In one, she substantiated that the pattern of tiny droplets on a pistol—which had subsequently been in a fire, rendering positive identification of the blood impossible—was consistent with "blowback"—blood spattering on the weapon at the time of firing, indicating that the pistol was very close to the victim when it was fired.[32] Bunker reported that "the locations of the stains and their size and shape are compatible with spatter produced by a gunshot wound."[33]

Another case in which Bunker assisted illustrates the value of crime-scene photographs in reconstructing the events in a questioned suicide. One of us (J.N.) was commissioned to reinvestigate the case for a bereaved family who believed that the death of their loved one was an accident. The deceased had been found dead in his apartment, lying at the foot of the stairs. He had been shot through the head, the bullet traversing the skull at a steeply *downward* angle but entering the wall, about mid-point of the stairway, at a sharply *upward* angle. Blood spatters recorded in the several official police color photographs enabled Bunker, working with a forensic pathologist, to reconstruct the victim's position when he was shot, a position consistent with his having fallen head-first down the stairs. (In addition, the photos yielded clues that had previously been overlooked, such as one of the victim's shoes being untied, providing a plausible cause for his fall, and the fact that he was wearing a cap when shot, confirming, as he had just said on the phone to his mother, that he was immediately leaving for work. His job required that he handle large sums of money, and he carried a loaded automatic pistol to work each day.) Bunker's conclusions (together with the findings of a "psychological autopsy"—an expert inquiry that seeks to determine whether someone was potentially suicidal—and other evidence) were enough to induce the state medical examiner's office to withdraw the determination of suicide.[34]

The process of conducting on-scene blood-pattern analysis is exacting work involving photographing and measuring each blood droplet and using a formula to calculate the impact angle. The criminalist then extends from each droplet a rod or string in the direction of its flight. The point at which the rods converge is the point where the blood originated.[35] According to one source, however, "suspending these strings or rods in the air is tedious, and it is easy to misinterpret the information rendered.

One odd stain pattern can throw off the entire reconstruction, and make the evidentiary conclusions less compatible." A forensic computer software program has been developed in an attempt to make the forensic expert's work easier, faster, and more reliable. The operator uses a special protractor to determine blood droplets' angles of travels and enters into the program, called "No More Strings," the length and width of each stain. The result is a simulated three-dimensional view of the area with lines drawn to indicate each stain's origin.[36] Depending on the particular circumstances, graphical aids, models, or videotaped recreations—in short, whatever may assist others in understanding the theory of events advanced by investigators—are used as aids for crime reconstruction.

Quite often the crime scene is no longer preserved when the need for a second (or third) opinion is realized. This was the situation in the case of Janet Cannon Myers's murder in New Orleans in 1984, the subject of Joesph Bosco's book *Blood Will Tell: A True Story of Deadly Obsession.* Judith Bunker was called in to review the evidence. From the photographs and blood stains collected at the crime scene, she was able to reach some definite conclusions; for example, that "Janet Myers received multiple blows to the head while lying on the living room floor. . . . This subject was turned to her left, head resting upon her left arm when trauma [was] inflicted to the right side of her head. Arterial spurting occurred as a result of this wounding, with blood projected onto the carpet and the *east* wall, *south* of the body. . . . This subject then turned to a supine position, her right arm in motion, forming a semicircular transfer pattern upon the carpet. Another blow was inflicted while her right arm was above her head."[37] But Bunker also pointed out problems resulting from the processing of the crime scene: "A foyer and den contain bloodstain evidence indicating the movement and directionality of people and objects during the following bloodshed. However, due to the absence of detailed photographs with scale and the limited number of stains collected for serological examination, a total reconstruction of the chain of events is not possible," she concluded.[38] Had the original investigators taken more care in documenting the crime scene, Bunker might have been able to provide a more detailed analysis.

LEGAL CONSIDERATIONS

In light of the travesty that was the O.J. Simpson trial in 1995, it appears that criminalists and investigators now have had an almost impossible burden added to their load: they must prove that they did not falsify

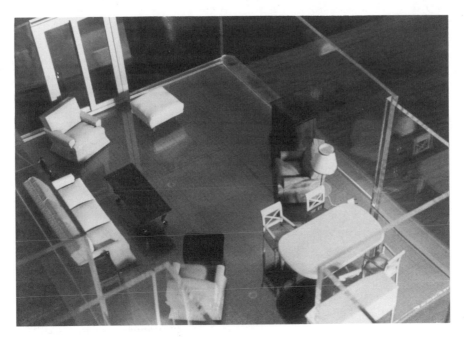

FIGURE 2.10. Scale models are frequently prepared for presentation to jurors in court. Crime-scene sketches, notes, and measurements prove vital for the subsequent preparation of such court models. (Courtesy of Don Ostermeyer, J.L. Bunker & Associates, Ocoee, Florida.)

evidence. Although laboratory analysis of the blood stains recovered from the murder scene, Simpson's Bronco, and his home tied him directly to the crime, Simpson's defense team countered with a battery of accusations. They not only accused the criminalists of mishandling and contaminating the evidence but also suggested that specific evidence—such as a bloody glove at Simpson's residence, the mate to one discovered at the scene—was planted. They insinuated that Detective Philip Vannatter took blood that a police nurse had collected from the suspect and sprinkled it at his home to incriminate the former football star.[39]

Los Angeles Police Department criminalist Dennis Fung conceded that evidence could possibly have been contaminated and that a white blanket used to cover the body of homicide victim Nicole Brown Simpson at the scene had been discarded before it could be examined for trace evidence such as hairs and fibers. Fung also was forced to admit that it was not he but a trainee who had collected most of the blood samples at the three locations.[40] Although the criminalists' and investigators' work was not nearly as faulty as suggested by the defense, clearly a number of mistakes were made and a number of procedures followed that, in hindsight at least, should not have been allowed.

Another issue that arose out of the Simpson case (which is the case study following chapter 8) is the matter of unreasonable search and seizure. The Fourth Amendment to the United States Constitution regulates the right of search and seizure: "The right of the people to be secure in their persons, houses, papers, and effects, against unreasonable searches and seizures, shall not be violated, and no warrants shall issue, but upon probable cause, supported by oath or affirmation, and particularly describing the place to be searched, and the persons or things to be seized."[41] "Probable cause" is defined by the U.S. Supreme Court as "a reasonable ground for belief, less than evidence justifying a conviction, but more than bare suspicion. Probable cause concerns circumstances in which a person of reasonable caution would believe an offense has been or is being committed."[42]

The Supreme Court has held that police may search without a warrant only in an emergency, or when securing evidence from imminent destruction, or with the consent of the parties involved, or when searching a lawfully arrested person and/or property within his immediate control. The Court has held that while in emergency circumstances, such as during a fire, initial entry and an accompanying search for evidence relating to the cause may be permitted, subsequent reentries without a warrant are invalid, and the evidence collected on those occasions is inadmissible.[43]

In the Simpson case, defense attorneys sought to have evidence obtained from Simpson's estate, notably the incriminating bloody glove, excluded because the detectives searched the premises before a search warrant was secured. Prosecutors countered that, after seeing blood traces on the Bronco and receiving no answer at the Simpson gate although lights were on inside the compound, detectives were responding to what they thought was an emergency situation; they argued that they were permitted to enter without a warrant and to search until a warrant was obtained in due course. At a hearing prior to the actual trial, municipal court judge Kathleen Kennedy-Powell ruled in the prosecution's favor.[44]

At all stages of endeavor the work of the criminalist must be circumspect. To prevent the exclusion in court of evidence on some real or perceived grounds, the documentation of a crime scene must be careful and accurate, and physical evidence must be collected thoroughly. As previously discussed, control samples must be taken whenever appropriate, and evidence must be preserved, kept free from contamination, marked properly, and logged consistently. The chain of custody must be rigidly maintained. Though, of course, hindsight affords the best vision, whenever a case is high-profile or *prima facie* complex, extraordinary care must be taken to provide the best evidence possible for other investigators' review, as the following case study—and many other cases—will attest.

CASE STUDY: THE JEFFREY MACDONALD CASE

At 3:40 A.M. on February 17, 1970, a telephone operator in Fayetteville, North Carolina, received an emergency call. The voice—very faint—was that of a man asking for an ambulance and military police to be dispatched to a local address.

"Is that on post or off post?" asked the operator. She received no answer.

At 3:42 A.M. a Fort Bragg desk sergeant got a call. He heard an urgent voice say, "Five forty-four Castle Drive—Help! Five forty-four Castle Drive! Stabbing!" He heard a clattering sound as if the caller had dropped the phone. After several seconds the caller was back; he repeated the latter part of his message and added, "Hurry!"

Ten minutes later a dozen or so military policemen were at the Castle Drive location. A sergeant found entry through the back door, which opened onto a little utility room that led to the master bedroom. What the sergeant saw there sent him running back to the front of the house, shouting, "Tell them to get Womack [Hospital] ASAP!"

MPs found Colette MacDonald, age twenty-six, lying on her back, drenched in blood. One eye was open. Her chest was partially covered by a ripped and bloody man's pajama top, one breast exposed. Her head had been badly battered. A paring knife lay nearby.

Lying motionless next to her was her husband, Green Beret Captain Jeffrey MacDonald, M.D., also twenty-six years old. He was wearing blue pajama bottoms that matched the top that covered his wife. A blood-soaked bedspread and sheet lay in a pile near the hall doorway. On the headboard of the couple's bed, large letters written in blood spelled "PIG." An MP heard MacDonald moan and rushed to his side. MacDonald whispered, "Check my kids. I heard my kids crying." Down the hall the MP found a bedroom and aimed his flashlight at the bed. He found five-year-old Kimberly MacDonald, the couple's eldest daughter, bludgeoned, with stab wounds in her neck. Across the hall, in another bedroom, was the body of MacDonald's youngest child, Kristen, age two. Blood had run over the bed sheets and pooled on the floor. Near this pool was a bloody footprint, leading back to the doorway.

Before the ambulance arrived, MacDonald told the MPs there had been four assailants, two white males, one black, and a female—all "goddamned acid heads." As the medics arrived, one of the MPs inside unlocked the front door and let them into the living room. They wheeled a stretcher through the room and down the hall to the master bedroom, where they placed MacDonald, the sole survivor, on it.

As they passed Kimberly's room on the way out, MacDonald suddenly grabbed the door casing and nearly pulled himself off the stretcher. "Goddamn MPs!" he shouted, as the soldiers restrained him. "Let me see my kids!" While a small crowd of neighbors milled outside, MacDonald was placed in the ambulance and taken to Womack Hospital.[45]

Doctors who attended MacDonald found that his wounds—which he would later exaggerate, claiming he had been stabbed twenty-three times—were relatively slight. He did have two slight knife cuts, a "neat, clean incision" between two ribs that had punctured a lung, a bump on his head, and a few fingernail scratches.[46]

Jeffrey MacDonald was soon interviewed by agents of the military's Criminal Investigation Division (CID) as well as by the FBI. He stated that he had stayed up to watch TV in the living room and to read a paperback book. About 2:00 A.M. he had washed the evening's dinner dishes. When he started to bed, he discovered that his younger daughter, Kristen, whom he'd earlier given a bottle of chocolate milk, was sleeping beside her mother and had wet the bed. He took her back to her own room and, to avoid waking his pregnant wife to change the sheets, took a blanket from Kristen's room and fell asleep on the couch.

In time, he was awakened by his wife's shouting from the bedroom, "Jeff! Jeff! Help!" and "Jeff! Jeff, why are they doing this to me!" Simultaneously, he heard his older daughter, Kimberly, screaming over and over, "Daddy! Daddy! Daddy!" As MacDonald opened his eyes, he claimed, he saw the four people he had earlier described to an MP, standing over him. The black man wore an Army fatigue jacket, and one of the two white men had on a sweatshirt with a red hood. The blond woman wore a floppy hat and high boots and carried a candle. All were dripping wet from the rain. The woman was chanting "Acid [the drug LSD] is groovy. Kill the pigs."

MacDonald said he tried to sit up but was struck with a baseball bat; in the ensuing struggle, he was stabbed with an icepick. He tried to wrestle his way to the hallway that led to his wife and children but passed out.

Later he awoke to silence and cold, his ripped and bloody pajama top twisted about his wrists. He went to each bedroom in turn, discovering the bodies of his wife and little girls, checking in vain for pulses and attempting mouth-to-mouth resuscitation. He heard only gurgling noises that indicated the escape of blood and air through the lungs. He removed the knife that was imbedded in his wife's chest and covered her with his pajama top. He went to the hall bathroom to check his own injuries, then telephoned for assistance.

Back at 544 Castle Drive, while MacDonald was being questioned at the hospital, CID agents were processing the crime scene. About four o'clock that morning, agent William Ivory had been on duty and had heard radio transmissions about the murders. He had grabbed evidence-processing equipment, summoned the CID photographer, and drove to the MacDonald residence. Ivory went briefly from room to room for a first look, switching on lights in the girls' bedrooms with the tip of his pen to avoid smudging any fingerprints. When the photographer arrived he became so nauseated at the bloody tragedy that he had to be escorted from the scene; he was soon replaced by the CID photo lab director himself.

Investigator Franz Joseph Grebner also arrived. Grebner studied the living room and was struck by the relative lack of disarray, amounting to only an upended coffee table and a spilled flowerpot. An adjacent dining area had not been disturbed at all. Grebner phoned the CID laboratory at Fort Gordon, Georgia, and requested that a team of crime-scene technicians be dispatched.

At daylight, agents found, just outside the back door, the murder weapons (except the paring knife MacDonald claimed he had removed from Colette's chest). The club had two blue threads stuck in the blood on it. Beneath a bush were the icepick and another knife. After the bodies had been photographed *in situ,* medics arrived about 8:00 A.M. to remove them. Ivory assisted by using tongs to remove items that had to be collected first, such as the pajama top that partially covered Colette's chest. After sealing the pajama top in a plastic bag, Ivory stood by as the medics lifted her body; he noticed, where her head had lain, a blue thread sticking out of a clot of blood.

After the medics left with the three bodies, Ivory continued to search the scene and collect evidence. Within the outline he had drawn on the bedroom rug around Colette's body, he discovered numerous small, one-to two-inch-long blue threads. Matching threads were scattered about the bedroom. Ivory also discovered in a corner of the room a finger section torn from a disposable surgical glove. It was stained as if it had been dipped in blood.

The CID laboratory team arrived and subsequently spent four days inside the MacDonald residence. They located a total of eighty-one blue threads in Jeffrey and Colette MacDonald's bedroom, an additional nineteen in Kimberly's, and two in Kristen's. Yet in the living room—where MacDonald claimed his pajama top had been ripped in struggling with the assailants—not a single thread was found, despite the fact that an evidence technician spent hours searching through the nap of the carpet with a magnifier.

Blood was found in profusion in the three bedrooms, including spatters high on the walls where it was apparently flung off the bloody club as it was raised. However, there were also locations notable for their *absence* of blood evidence. For example, none was found on the hall floor where MacDonald stated he had lain unconscious. No blood—or fingerprints—were found on either of the two telephones he had used to call for help. The same was true for the living room, with two exceptions: a speck of blood on the outside of one lens of a pair of MacDonald's eyeglasses, found in a corner, and a smear, as from a fingertip, on the upper edge of a magazine. Several drops of blood were found on the bathroom sink, and five more drops were discovered on the kitchen floor, directly in front of a cabinet that contained several packages of disposable latex surgical gloves.

The word "PIG" formed in blood on the headboard of the MacDonald's bed was studied carefully. It appeared to have been written by a right handed person using the first two fingers enclosed in something similar to a latex glove. Beneath the headboard were found two of the eighty-one blue threads discovered in the bedroom.

In the meantime, autopsies had been conducted on the murder victims in the hospital's basement morgue. Colette had been stabbed with a knife a total of sixteen times, nine in the neck and seven in the chest, and an additional twenty-one times in the chest with an icepick driven in up to the hilt. She had been bludgeoned in the head six times, and both of her arms were broken in what the pathologist called "defensive type injuries" sustained while she had held her arms up in an attempt to shield herself from the blows. In addition to her pregancy, the autopsy revealed a piece of what appeared to be human skin under one of her fingernails.

Kimberly had been hit on the head at least three times. She was also stabbed in the neck repeatedly. Kristen had multiple stab wounds—twelve in the back, four in the chest, and one in the neck, with an additional fifteen shallow wounds in the chest that appeared to have been caused by an icepick. Several cuts on her hands were also, clearly, defensive wounds. At the CID laboratory, serology experts learned the unusual fact that all four family members had a different blood type: Colette's was A, her husband's was B, Kimberly's was AB, and Kristen's was O. This knowledge enabled investigators, who tested each of the myriad stains, to reconstruct the essential events of the fatal morning.

Unfortunately, blunders were made at the crime scene and at the laboratory. When a technician attempted to use a saw to remove the bloody footprint on the floor of Kristen's bedroom, the boards separated and the print was destroyed. Had photographs been taken of the print before the saw was used, the evidence would at least have been documented in that form. Unaccountably, at least two pieces of collected evidence—the piece of skin from beneath Colette MacDonald's fingernail and a blue fiber from beneath Kristen's fingernail—were lost.[47] Had proper handling, storage, and logging procedures been followed, that evidence would have been available for use in criminal proceedings.

Six weeks passed, during which time Grebner became increasingly skeptical of MacDonald's account. Finally, on April 6 he summoned the Green Beret captain to CID headquarters for further interrogation. MacDonald was advised of his Miranda rights.[48] A tape recorder preserved his acknowledgment that he understood those rights and that he would answer any questions that were put to him. With the recorder running, Grebner said to MacDonald, "Just go ahead and tell us your story." MacDonald cleared his throat and repeated his account of the four intruders—this time on record.

The recorded interrogation provided an account that investigators could then compare with their reconstruction of the crime. Unfortunately for Capt. Jeffrey MacDonald, the physical evidence was at considerable variance with his story in several major respects. The evidence at the scene—specifically the lack of blood in appropriate areas—did not support his story of being stabbed in the living room and passing out in the nearby area of the hallway. The footprint in Kristen's room, made with Colette's blood, was one of several indications that she was first bludgeoned there. The blood in front of the kitchen cabinet where the rubber gloves were kept was type B, Jeffrey MacDonald's type.

Many additional findings pointed to MacDonald as the killer, and none pointed to four intruders. MacDonald had lied about not owning an icepick. Investigation revealed he had had several extramarital affairs. No physical evidence indicated that anyone else had been in the house at the time of the murders. That four intruders could enter the house, murder three people and assault a fourth, leave the murder weapons at the scene and yet leave no physical evidence of their presence was nearly impossible.

Following MacDonald's interrogation, during which he offered to take a polygraph test but later withdrew the offer, he was informed he was a suspect in the case. He was relieved of duty and immediately placed under restriction.

Three months later, a military hearing was held. Such a hearing ordinarily would have been perfunctory, intended only to determine whether the charges had sufficient merit to warrant a court-martial. However, MacDonald attorney Bernie Segal determined to put on a full defense; the prosecution was unprepared. Segal attacked vigorously, making much of the mishandling and loss of evidence and accusing the CID of failing to investigate adequately the alibi of one Helena Stoeckley, who supposedly owned a blond wig and floppy hat, was given to extensive drug use, and could not recall her activities on the night of the murders. (CID investigators had questioned Stoeckly and had found no reason to believe she had had anything to do with the crimes.) Segal put on witnesses who testified as to MacDonald's character and described the love they believed he had for his family. MacDonald had favorably impressed a panel of military psychiatrists, the chief of which testified at the hearing as to MacDonald's soundness of mind and apparent sincerity. Finally, Segal called MacDonald himself to testify, and for three days he impressed the hearing's presiding officer, Col. Warren V. Rock, who ultimately recommended that the charges against MacDonald be dismissed because they were "not true" and that Helena Stoeckley's activities and whereabouts be investigated as to "the early morning hours of 17 February 1970." The Army chose, however, not to make the colonel's recommendations public and simply announced that the charges against MacDonald were dismissed for "insufficient evidence." He subsequently sought and obtained an honorable discharge.

During all of this, Colette's mother and stepfather, Mildred and Freddy Kassab, were Jeffrey MacDonald's greatest defenders. Over time that began to change. Freddy Kassab was upset with his son-in-law's grandstanding appearance on *The Dick Cavett Show,* and he came to see that they had quite different views about the murders. Kassab wanted to pursue the case and bring the criminals to justice, but MacDonald seemed to want to put the matter behind him and get on with his life—and his womanizing. Kassab had not attended the hearing and was surprised to learn, from MacDonald's stalling and outright lying, that he did not wish Kassab to know details of the evidence. At one point, to forestall Kassab's pursuit of the matter, MacDonald actually claimed to have tracked down and killed one of the four murderers. When Kassab finally obtained a transcript of the hearing and talked with the CID investigators, he made a complete turnaround and sought to have MacDonald brought to trial.

Eventually it fell to Justice Department attorney Victor Woerheide to determine whether the case was worthy of prosecution. His review persuaded him in the affirmative, and he ordered a federal grand jury convened in Raleigh, North Carolina. For five days in August 1974, four years after the murders, MacDonald testified before the grand jury. Other witnesses included an Army clinical psychologist who had reviewed previous tests of the suspect, and Paul Stombaugh, chief of the FBI laboratory's chemistry section, who had analyzed the crime-scene evidence in 1971 and more recently had conducted further analyses on the various bloodstained garments and other evidence. He found that Colette's body had been covered with a sheet and moved by someone wearing Dr. MacDonald's blue pajama top, which left a bloodstained imprint of its sleeve on the sheet; that some of Colette's blood had stained the blue pajama top *before* it was torn; that the club, which MacDonald denied knowledge of, had been sawed from the same board as one of the mattress slats; that Colette was stabbed not with the knife MacDonald claimed he had pulled from her chest but by another knife, which had been found under the bush at the rear of the house; and that the forty-eight holes in MacDonald's blue pajama top had all been made when the top was stationary rather than in motion as it would have been if the punctures were made during a struggle with intruders.[49]

On January 21, 1975, MacDonald was recalled to appear before the grand jury to answer further questions. An indictment was subsequently handed down. MacDonald's attorney, Bernie Segal responded by filing a motion claiming the indictment constituted double jeopardy and that MacDonald had been denied the right to a speedy trial. A federal district court judge denied both motions, ruling that the previous hearing was not the equivalent of a trial, and that the right to a speedy trial did not apply until after an indictment had been returned. Segal appealed, whereupon the Fourth Circuit Court of Appeals ordered dismissal of the indictment. Finally, after a petition to the U.S. Supreme Court and a delay of another year, the Court announced it would hear the case in January 1978. Finally, the Court ruled that the appellate court could not act on a speedy trial motion before trial; doing so, the Court held, would only "exacerbate pretrial delay."

In 1975 Justice Department attorney Victor Woerheide had suffered a fatal heart attack, and James L. Blackburn had been appointed in his place. Now, almost nine and a half years after the murders, Jeffrey MacDonald was brought to trial. It began in mid-July 1979 after three days of jury selection. FBI expert Stombaugh's testimony proved crucial and was fiercely debated, especially his laboratory demonstration that MacDonald's blue pajama top, when refolded as nearly as possible to its appearance in the crime-scene photos, had holes that were consistent in alignment with the icepick wounds in Colette's body—sixteen on the left and five on the right. Segal tried repeatedly to suppress the evidence and to discredit Stombaugh—even to the point of making fun of his use of a magnifying glass, which Segal argued was "not a scientific instrument."[50]

During cross-examination of MacDonald, Jim Blackburn went over each point of evidence against him, asking him if he was able to explain each in turn. Many of the defendant's responses were ineffectual, and he frequently claimed that he had no idea about a particular point of evidence. After closing arguments, the seven-week trial went to the jury. They took just six and a half hours to reach a verdict. MacDonald was found guilty of second-degree murder in the death of his wife and Kimberly, who were judged to have been killed in a rage, and guilty in the first degree in the death of Kristen, who evidently was killed only to lend credence to his cover story. He was sentenced to three consecutive life sentences. Twice the appeals went to the Supreme Court, which decided on March 31, 1982, that MacDonald's right to a speedy trial was not violated and declined on January 10, 1983, to review the court of appeal's rejection of the claim that MacDonald had failed to receive a fair trial.[51]

Just three days after the Supreme Court rendered its final decision, on January 13, 1983, Helena Stoeckley died. Influenced by promises of financial assistance and help in starting a new life, Stoeckley had signed a statement implicating herself and five friends in the murder of the MacDonald family. She later recanted, saying she was "a pawn" who had been "coerced" into signing a "so-called 'confession'" and that she had been "exploited by means of false hopes and empty promises." Then she recanted this recantation and confessed yet again on the television program *Jack Anderson Confidential*. Obviously, she became a classic example of "false memory syndrome."[52] Neither her hair nor her fingerprints matched any evidence discovered at the murder scene.[53]

Despite blunders that resulted in lost evidence, the processing of the crime scene permitted the reconstruction of the crimes, and that in turn eventually brought justice to the victims who died at 544 Castle Drive.

MacDonald's address is now the Federal Correctional Institution at Bastrop, Texas.

After having examined the physical evidence in the case at the request of the defense and being asked to testify for MacDonald, blood-pattern analyst Judith Bunker replied, "No. My testimony would only hurt your client."[54]

NOTES

1. Charles E. O'Hara, *Fundamentals of Criminal Investigation,* 3d ed. (Springfield, Ill.: Charles C. Thomas, 1973), 47.

2. This list of procedures is taken primarily from ibid., 47-48; see also Vernon J. Geberth, *Practical Homicide Investigation,* 2d ed. (Boca Raton, Fla.: CRC, 1993), 37-39.

3. Geberth, *Practical Homicide Investigation,* 2.

4. Charles R. Swanson Jr., Neil C. Chamelin, and Leonard Territo, *Criminal Investigation,* 4th ed. (New York: McGraw-Hill, 1988), 37-38.

5. O'Hara, *Fundamentals,* 48-49.

6. Fred E. Inbau, Andre A. Moessens, and Louis R. Vitullo, *Scientific Police Investigation* (New York: Chilton, 1972), 6-7.

7. Swanson, Chamelin, and Territo, *Criminal Investigation,* 52.

8. Ibid., 52-53.

9. Ibid., 54; Inbau, Moenssens, and Vitullo, *Scientific Police Investigation,* 7-8.

10. Swanson, Chamelin, and Territo, *Criminal Investigation,* 53-54.

11. R.M. Boyd, "Buried Body Cases," *FBI Law Enforcement Bulletin,* vol. 48, no. 2 (Feb. 1979): 1-7, cited in ibid., 256-61.

12. O'Hara, *Fundamentals,* 64; Swanson, Chamelin, and Territo, *Criminal Investigation,* 48.

13. O'Hara, *Fundamentals,* 64-65.

14. Ibid., 67-68.

15. Ibid., 66-67. Swanson, Chamelin, and Territo, *Criminal Investigation,* 50; illus., 53.

16. Swanson, Chamelin, and Territo, *Criminal Investigation,* 48-50.

17. Ibid., 39; O'Hara, *Fundamentals,* 49; Geberth, *Practical Homicide Investigation,* 93; Richard Saferstein, *Criminalistics: An Introduction to Forensic Science,* 5th ed. Englewood Cliffs, N.J.: Prentice Hall, 1995), 35.

18. Saferstein, *Criminalistics,* 35.

19. Ibid. 35-36.

20. Ibid., 138-39; O'Hara, *Fundamentals,* 50-52.

21. As Chief August Vollmer long ago said of professional criminals, "Not only do they specialize in particular crimes, but within the ranks of specialists we find difference in their mode of attack which are useful hints to careful investigators." Quoted in "The Modus Operandi System," instruction manual for course

in Scientific Crime Detection (Chicago: Institute of Applied Science, 1962), lesson 1, 4.

22. O'Hara, *Fundamentals,* 50.

23. Ibid., 77.

24. Ibid., 78-80.

25. See Ordway Hilton, *Scientific Examination of Questioned Documents,* rev. ed. (New York: Elsevier Science, 1982), 353. In the field of questioned documents the tendency is now to avoid marking the document and instead to rely on accurate notes.

26. O'Hara, *Fundamentals,* 80-81.

27. Ibid., 74-75.

28. Ibid., 55-56.

29. Ibid., 56.

30. Timothy M. Dees, "Simplifying Blood Spatter Analysis at the Crime Scene," *Law Enforcement Technology* (Aug. 1995): 42-44.

31. Barry A.J. Fisher, *Techniques of Crime Scene Investigation,* 5th ed. (New York; Elsevier, 1992), 218-23.

32. Joe Nickell with John F. Fischer, "The Case of the Shrinking Bullet," in *Mysterious Realms: Probing Paranormal, Historical, and Forensic Enigmas* (Buffalo, New York: Prometheus, 1992), 107-29.

33. Judith Bunker's report, quoted in ibid., 125.

34. Abridged from Joe Nickell, *Camera Clues: A Handbook for Photographic Investigation* (Lexington: Univ. Press of Kentucky, 1994), 99-100. The cause of death was revised to "cannot be determined."

35. Dees, "Simplifying Blood Spatter-Analysis," 42-43.

36. Ibid., 43-44.

37. Joseph Bosco, *Blood will Tell: A True Story of Deadly Obsession* (New York: William Morrow, 1993), 131.

38. Ibid.

39. Clifford Linedecker, *O.J. A to Z: The Complete Handbook of the Trial of the Century* (New York: St. Martin's Griffin, 1995), 31-32, 238-39.

40. Ibid., 95.

41. The Constitution of the United States, Amendment IV.

42. Quoted in Russell Bintliff, *Police Procedural: A Writer's Guide to the Police and How they Work* (Cincinnati: Writer's Digest, 1993), 58.

43. Supreme Court, *Michigan v. Tyler,* 56 L. Ed. 2d 486 (1978), cited in Saferstein, *Criminalistics,* 49-50.

44. Bintliff, *Police Procedural,* 82-83.

45. This account is taken from Joe McGinnis's *Fatal Vision,* updated edition (New York: Signet, 1989), 11-16.

46. Ibid., 20, 583.

47. Ibid. 35-38, 58-60, 96-98.

48. Established in a 1966 Supreme Court ruling, in *Miranda v. Arizona;* Bintliff, *Police Procedural,* 202. MacDonald was informed that he had the right to an attorney before further questioning and that what he said could be used against him.

49. McGinnis, *Fatal Vision,* 96-100, 107-160, 176-243, 264-505.

50. Ibid., 505.

51. Ibid., 501-659.

52. For a discussion of this syndrome, see Elizabeth Loftus, "Remembering Dangerously," *Skeptical Inquirer,* vol. 19, no. 2 (Mar./Apr. 1995), 20-29.

53. McGinnis, *Fatal Vision,* 241, 622-38. For a contrary view see Laurie P. Cohen, "Strand of Evidence," *Wall Street Journal,* April 16, 1997.

54. Bosco, *Blood Will Tell,* 130.

RECOMMENDED READING

Bosco, Joseph. *Blood Will Tell: A True Story of Deadly Obsession.* New York: William Morrow, 1993. Popular, true-crime account of a complex case that centers on blood pattern evidence and its analysis.

Dees, Timothy M. "Simplifying Blood Spatter Analysis at the Crime Scene." *Law Enforcement Technology* (Aug. 1995). Discussion of new computer program for analyzing blood-pattern evidence.

Fisher, Barry A.J. *Techniques of Crime Scene Investigation.* 5th ed. New York: Elsevier, 1992. Comprehensive textbook on all aspects of crime-scene securing, documenting, and processing; includes chapters on major categories of physical evidence such as firearms examination.

MacDonnell, Herbert Leon. *Bloodstain Patterns,* rev. ed. Corning, N.Y.: Laboratory of Forensic Science, 1997. Forensic treatise on blood pattern analysis by a pioneer in the field.

McGinnis, Joe. *Fatal Vision.* Updated ed. New York: Signet, 1989. Masterful account of the Jeffrey MacDonald case, illustrating the value of physical evidence from the crime scene; the source for the case study in this chapter.

O'Hara, Charles E. *Fundamentals of Criminal Investigation.* 3d ed. Springfield, Ill.: Charles C. Thomas, 1973. Includes basics of crime-scene documenting and processing in four chapters: "Crime Scene Search," "Photographing the Crime Scene," "Crime Scene Sketch," and "Care of Evidence."

Potter, Jerry Allen, and Fred Bost. *Fatal Justice: Reinvestigating the MacDonald Murders.* New York: Norton, 1995. Attempt to rebut McGinnis's *Fatal Vision* by challenging the physical evidence and witness testimony.

3 | TRACE EVIDENCE

The term *trace evidence* is a generic one, referring to minute physical evidence that may be transferred from a criminal to a victim or crime scene, or vice versa (Locard's Exchange Principle, discussed in chapter 1). Consider, for example, the body of a nude female dumped in a remote area: Fibers on the body may be traced to the seat covering or trunk carpet of a particular type of automobile; combings of head and pubic hair (with samples of the victim's hair taken as controls) may reveal hairs that are valuable in identifying a rapist; fingernail scrapings may not only indicate that the suspect is suspiciously marked but may also provide sufficient DNA samples for a conclusive identification. The most common types of trace evidence are *hair and fibers, soil and botanicals,* and *glass, paint, and other trace evidence.*

THE MICROSCOPE

The microscope (from the Greek words for "small" and "to look at") in its simplest form is a device with a single convex lens, such as a jeweler's loupe, a fingerprint expert's tripod magnifier, or an ordinary magnifying glass.[1] When most people use the term *microscope,* however, they are referring not to the simple but to the compound microscope, an instrument composed of two lenses—the objective lens and the ocular lens, or eyepiece—mounted in opposite ends of a tube.

The first such instrument appears to have been made by Zacharias Janssen about 1590.[2] Early microscopes of this type performed poorly, however, because of certain lens aberrations, particularly those chromatic in nature, such as color distortions caused by the glass failing to refract, or bend, all wavelengths of light equally. The two-stage magnification process accentuated these and other distortions. The so-called "first microscope," invented in 1673 by Anton van Leeuwenhoek, a Dutch naturalist, avoided these problems partly because it was actually a simple single-lens microscope, one that gained power by using a tiny, highly curved lens that was viewed through an opening only a little larger than a pinhole. Leeuwenhoek's invention largely eliminated the distracting "color fringes" around objects. Eventually the problem of chromatic aberration was solved by using different types of glass with different refractive indices for each lens, creating *achromatic* lenses.[3]

Powerful magnification sometimes can be a hindrance rather than an assistance. Just as one may fail to see the forest for the trees, one may fail to see the tree for its leaves or the leaf for its microscopic cells. Generally speaking, the larger the magnification, the smaller the area that can be viewed at one time. So, an ordinary magnifying lens, or "reading" glass, with its approximately two- to four-power magnification and its relatively large field of view, is appropriate for many first-look forensic purposes, such as searching the crime scene. Magnifiers and loupes, usually with double lenses or a single thick lens, are convenient for many kinds of fieldwork. A ten-power Bausch & Lomb "illuminated Coddington" magnifier features a single thick lens with penlight illumination through a central groove that provides an exceptionally clear, bright image—especially when used not like a magnifying glass but as a loupe, held against the eye.[4] When higher magnification is required, microscopes are used, and various types are suitable for different forensic applications.

Standard Laboratory Microscope. This is the microscope the average person is most familiar with, sometimes called a "biological" microscope because of its many applications in medicine and related fields. It is designed almost exclusively for transmitted light illumination in which a very thin specimen is placed on a glass slide and light transmitted through it. This type of microscope is used in serological examinations and for other specific forensic uses. In a modified form it becomes the polarized light microscope.[5]

Polarized Light Microscope. Also known as the "polarizing" microscope, this instrument "is the most versatile and powerful tool available for analyzing trace evidence."[6] It is essentially a standard laboratory microscope to which certain refinements have been made, notably the addition of polarizing light filters. (One filter transmits light vibrating in one direction, front to back, while the other transmits light in a left-to-right direction.) These create a field that appears dark, but certain materials, including many crystalline substances, have distinctive properties of *birefringence* (light passing through them is doubly refracted), making them easily identifiable under polarized light. (Because this method is the best for studying the optical properties of minerals, it is used by geologists, who call it the "petrographic" microscope.) Forensic experts use it for examining minerals in soil, synthetic fibers, glass, and other birefringent materials.[7]

Stereoscopic Microscope. This, the workhorse of the crime lab, is the most versatile and widely used instrument in the criminalist's arsenal. This type of microscope actually consists of two side-by-side, relatively low-power compound microscopes in a common housing. They are placed so that each is directed at the same specimen area from a slightly different angle. When used with a binocular eyepiece, there is a microscope for each eye, creating true "stereoscopic" viewing.[8] According to one authority, "The fact that microscopic objects tend to have very natural 'three-dimensional' appearances when viewed with the stereomicroscope is the instrument's chief advantage. It is this attribute that allows the stereomicroscope to be thought of as a direct extension of normal vision into the microscopic realm. For this reason, little training is required in order for an examiner to use this instrument productively."[9]

With the stereomicroscope, comparatively low magnification (usually ten- to sixty-power) and direct light (either reflected or oblique) are used. This permits the viewing of relatively large objects, such as mineral specimens or—by transferal of the microscope's body from its usual base to one with an adjustable extension arm—large documents or other items with a sizable surface area. Its stereo or "3-D" image enables a document examiner to accurately view such subtle, depth-related features as nib tracks (furrows in paper left by steel or other hard-nibbed pens), crossed strokes, and erasures on paper. Used in every branch of criminalistics, it is frequently the first instrument applied in preliminary evaluation of evidence or for sorting or selecting items for further examination.[10]

Comparison Microscope. This is essentially a pair of precisely similar compound microscopes so arranged that when a viewer looks through the eyepiece the two images are juxtaposed in a common field plane. (Some models also permit the images to be superimposed as well.) This allows two separate objects—such as two bullets or two shell cases—to be directly compared. Not only were forensic problems among the earliest

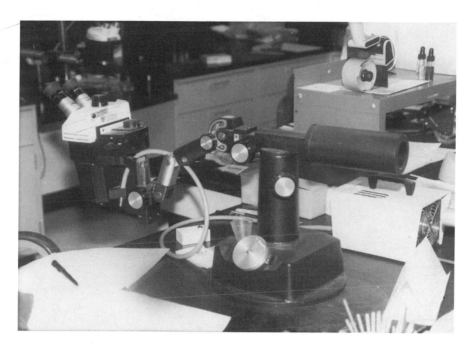

FIGURE 3.1. Probably the most frequently used microscope in the crime laboratory is the stereomiscroscope. Samples of trace evidence such as vacuum sweepings recovered from crime scenes are generally examined first with this instrument. (Courtesy of Robert Miller, Phoenix Police Department, Phoenix, Arizona.)

applications of the comparison microscope, but early firearms expert Calvin Goddard employed the instrument specifically to refine the process of making firearms comparisons.

In addition to this type of comparison microscope, which, like the stereomicroscope, utilizes low power and direct lighting to examine opaque objects, there is a second major type that is essentially a combination of two standard or polarized microscopes. This type of comparison microscope therefore uses transmitted light and is ideal for comparing hairs and fibers.[11]

Other Microscopes. In addition to the basic instruments just described, there are much more exotic types of microscopes, including the scanning electron microscope (SEM), which is capable of a wide range of magnifications from relatively small (15x) to exceedingly large (over 50,000x). Invented in 1938, it is based on the principle that, like light, electrons are reflected from whatever they strike. The electrons reflected by a target object are picked up by a magnetic field and converged upon

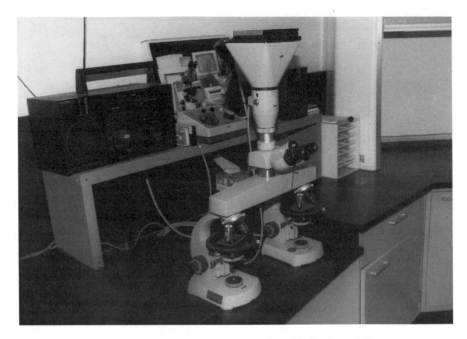

FIGURE 3.2. A comparison microscope specifically designed for examination of trace evidence is generally used to compare questioned hairs and fibers with known standards. This comparison microscope has a camera attachment. (Courtesy of Robert Miller, Phoenix Police Department, Phoenix, Arizona.)

a chemically treated glass. The image may then be projected—and enlarged—on a phosphorescent screen. Most SEM pictures have a remarkably clear, three-dimensional appearance, thus making the SEM an excellent instrument for examining small traces of material such as paint and gunshot residues.[12] However, one forensic authority cautions:

> When using these powerful instruments in forensic work, it is well to keep in mind that no two objects are ever exactly the same. No two sand grains are ever exactly alike when studied under the high magnifications of an electron microscope. This is true even if they have been side by side for the past million years. Observations made with these instruments can be very useful for establishing comparison or lack of comparison between samples. However, the very power of the instruments permits the possibility of their abuse in the hands of the unscrupulous. If we were to do a complete chemical analysis of a total person by the most modern methods in the morning and repeat the analysis on the same person in the afternoon, we would find chemical differences. However, this would not demonstrate that we had analyzed two

different people. Thus, the demonstration of small differences in soil does not provide in itself that they do not compare. It is equally true that showing a common similarity among soil samples, such as their both containing quartz, is poor evidence on which to base a comparison. The professional judgment of the scientist thus becomes increasingly important when these powerful instruments are used.[13]

Another sophisticated instrument is the microspectrophotometer, which is actually two instruments in one: a standard laboratory microscope combined with a computerized spectrophotometer (an instrument that helps identify a substance by directing a beam of light at it and thus obtaining what is called its *absorption spectrum*). The limitation of the spectrophotometer is that it is poorly suited to the examination of tiny particles, whereas the microspectrophotometer makes that task relatively easy. Moreover, in addition to analysis using ordinary light, infrared radiation (IR) can be employed in order to obtain yet additional information about a particle in question. Notes one authority: "The 'fingerprint' IR spectrum . . . is unique for each chemical substance. Therefore, if such a spectrum can be obtained from either a fiber or paint chip, it will allow the analyst to better identify and compare the type of chemicals from which these materials are manufactured."[14]

Two techniques that are often important in examining microtraces are microchemical tests and photomicrography. Microchemical tests are accomplished by using a pipette to add a tiny amount of reagent (a test chemical) to the questioned material on the microscope slide. (For example, iodine as a reagent turns starch grains blue.) The reaction, if any, is observed through the microscope; this technique is useful for testing very small amounts of material. It can also confirm visual identifications.[15] The other technique, photomicrography (taking photographs through a microscope), permits the criminalist to document work. (For example, a particle may be photographed before adding a reagent, which may alter or destroy it.) Enlarged photomicrographs can also be used as courtroom exhibits.

HAIR AND FIBERS

Hair and fibers are superficially similar. They may frequently be found together as trace evidence, and the methods of studying them are often similar. But, in fact, hair and fibers are quite different: Hair is an appendage of human or animal skin, growing out of an organ known as

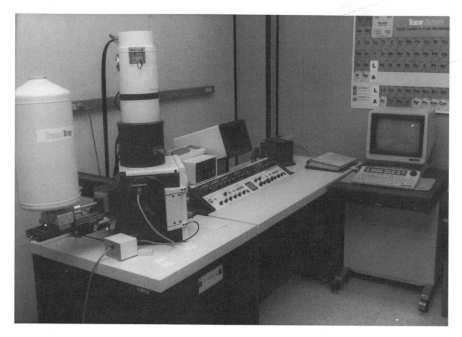

FIGURE 3.3. Scanning electron microscopes like this one are frequently equipped with x-ray analyzers that allow the magnified particles to be identified by their specific x-ray emissions. Gunshot residue particles are often examined with this type of microscope. (Courtesy of Robert Miller, Phoenix Police Department, Phoenix, Arizona.)

the *hair follicle*; fibers are strands of natural or manufactured material. To confuse matters, some animal-hair coverings such as wool and mohair are classified as natural fibers, but such issues of terminology are just that and do not affect the process of identification.

Hair. Although hair is discovered as evidence at many crime scenes and often appears on the bodies of victims of violence, it represents *class,* or *general,* rather than *individual* evidence. One person's hair cannot yet be individualized to the exclusion of anyone else, although DNA research, which involves testing of the *root* of the hair, may make near-individualization possible. In any event, hair is valuable evidence; much can be learned from a few strands. Fortunately, it resists chemical decomposition and retains its structural features for a long time. These features include the *cuticle,* the scaly exterior portion that is sufficiently distinctive to make it helpful for identification by species (human as opposed to canine, for example); the *cortex,* an intermediate layer that gives hair its color; and the *medulla,* a central canal that may be classified as continuous, interrupted, fragmented, or entirely absent.[16] A noted forensic expert on hair makes this interesting analogy:

> An ordinary lead pencil is an excellent model of the various structural components of hair. A pencil has a measurable length and diameter. The metal sheath tightly gripping the bulbous rubber eraser can be compared to the root structure of the hair. The wooden portion of the pencil represents the keratinized tissue. This wooden cortex often contains brown resin flecks representing pigment granules. The pencil's painted outer surface corresponds to the cuticular layer of the hair. A clear paint would better represent the cuticle, as the human cuticle is colorless. The black center core of the pencil which contains a graphite-like material, represents the medulla. The pencil point tip may be cut sharply, tapered, or rounded with use. Like hair, if the pencil is not cared for, the tip sometimes becomes broken or frayed.[17]

Allowing for biological overlap and racial variation, and within some limits, human hair may reveal the donor's race; the part of the body that yielded the hair; whether the hair was removed forcibly; whether it was cut and, if so, by what kind of instrument; whether it has been dyed or bleached; and whether it was subjected to some force. A tentative judgment as to sex and age of the donor (the latter within broad categories) may sometimes be rendered. Laboratory analysis may yield information about certain drugs or chemicals that may have been ingested; since hair grows about one millimeter per day, it may even be possible to draw conclusions about the last time a drug was ingested by analyzing different segments of a single donor hair. Still other trace evidence may be found adhering to hairs, including blood, various types of dust, minute fibers, and paint.[18]

In the examination of known and questioned hair samples in the laboratory, the comparison microscope (of the transmitted-light variety) is an essential instrument. The examiner considers a number of characteristics of human head hair when making a comparison.[19]

COLOR	white, blonde, light brown, brown, gray brown, dark brown, gray, black, auburn, red
REFLECTIVITY	opaque, gray, translucent, transparent, auburn, clear
LENGTH	fragment, 1", 1-3", 3-5", 5-8", 8-12", 12-18", 18-30", segment
DIAMETER	20-30 μm, 30-40 μm, 40-50 μm, 50-60 μm, 60-70 μm, 70-80 μm, 80-90 μm, 90-100 μm, 100-110 μm

SPACIAL CONFIGURATION	undulating, kinky, curly, wavy, curved, straight, sinuous
TIP	singed, uncut, tapered, rounded, cut 90°, cut at angle, frayed, split, smashed, broken
BASE	cut, damaged, pigmented, clear, enlarged, putrid, tapering, broken
ROOT	stretched, absent, bulbous, sheathed, atrophied, germ, follicular, wrenched, putrid
CROSS SECTION	polygonal, ribbon, flat, flat oval, oval, round oval, undulating, round, convoluted
PIGMENT	absent, liquid, nongranular, granular, multicolor, chain, massive (clumped), dense, streaked, opaque
MEDULLA	absent, sparse, scanty, fractional, broken, globular, continuous, irregular, double, cellular
CORTICAL FUSI	absent, few, abundant, bunched, linear, central, periphery, roots
CORTICAL CELLS	brittle, damaged, fibrous, cellular, invisible, fusiform, ovoid bodies
BIREFRINGENCE	gold, bright colors, dull colors, brown
COSMETIC TREATMENT	sun bleached, bleached, rinsed, natural, dyed, damaged
CUTICLE	ragged, serrated, looped, narrow, layered, wide, cracked, absent, clear, dyed
SCALES	flattened, smooth, level, arched, prominent, serrated

After human hair analysis, usually one of three findings can be expected: (1) the questioned sample is consistent with a known one in terms of microscopic characteristics and therefore either originated from the same individual or from another whose hair has the same qualities; (2) the samples are *not* similar in their characteristics and therefore did not come from the same person; or (3) the examiner is unable to reach a conclusion.[20] Limited though these are, it should be kept in mind that "a negative finding may serve to disprove erroneous theories. A positive finding, although merely suggesting the implication of an individual, is of importance when it is correlated with other newly discovered facts."[21]

As indicated in the previous chapter, hair discovered at a crime scene should be collected carefully by using tweezers or forceps and should be placed in bottles with snap-on tops. In rape cases, *medical personnel* should collect the samples, using a clean comb to remove any foreign hair from the pubic area and taking samples from the victim for controls. One

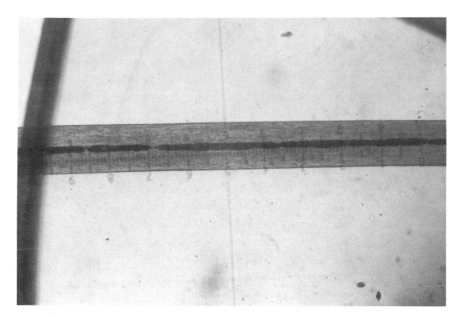

FIGURE 3.4. Photomicrograph of a Caucasian scalp hair (100x)

should keep in mind that hair differs in different parts of the body, so it is necessary to remove samples from each area that may be relevant to a case—especially the head and pubic areas with which most forensic comparisons of hair are involved. Forensic authorities disagree over how many strands should be collected, but somewhere between thirty and fifty from each location should be sufficient. The strands from different body areas—obtained by combing, or cutting if necessary—should be packaged separately and properly labeled.[22]

Fibers. In contrast to hair, fibers and cloth fragments offer much greater evidential value because they incorporate numerous variables. These include the number of fibers in each strand, the diameter of strands and fibers, the direction and number of twists, the type of weave, and the dye content, as well as the presence of any foreign material that may be adhered to fibers or embedded among them.[23]

Fibers may be classified into two rather broad groups—natural and manufactured. Until this century the former predominated, but, beginning with rayon, introduced in 1911, and nylon, developed in 1939, manmade fibers have largely come to replace natural ones in textiles. *Natural fibers* include animal (wool, silk, and animal-hair fibers such as mohair), vegetable (seed fibers such as cotton; bast fibers such as flax, hemp, and jute; and leaf fibers, including manila), and mineral (asbestos). *Man-made fibers* include natural-polymer fibers (rubber, rayon, and cellulose ester), synthetic-polymer fibers (polyvinyl derivatives, polyurethane, and polyester), and other fibers (carbon, glass, metal, and ceramic).[24]

The identification of natural fibers ranges from simple to much more complex procedures. Fibers from the hair coverings of sheep, goats, and other animals, and fur fibers from mink, beaver, rabbit, and so on are identified by their color and structural characteristics under the microscope. (Examination of a number of control samples is always necessary, of course, to determine the range of features that make up the questioned material.) Plant fibers also have distinctive morphologies, or structures, under the microscope that make them generally easy to identify. For example, cotton fibers have a characteristic twisted, ribbonlike form, and linen fibers resemble slightly knobby tubes. Cotton is the most prevalent of the plant fibers, so, alone, its evidential value is weak, but the presence of dye adds a significant measure of distinctiveness.[25]

For all but the most commonly found natural fibers, specialized techniques may be required. Maceration is necessary for examining bast and leaf fibers that occur in bundles. Cross-sectioning is employed to look for distinctive cross-sectional shapes. Obtaining casts, or impressions of the exterior scale pattern of animal fibers, allows identification of different species, and using polarized light microscopy determines birefringence.[26]

Man-made fibers generally show fewer identifying features than natural fibers, so several methods of identification are usually employed. These include infrared spectroscopy, which is valuable in differentiating the several types of nylon fibers; solubility tests to differentiate between certain other fiber types; and cross-sectioning. In addition, standard microscopical examination is used to determine optical properties, such as birefringence and refractive indices, which are determinable with the polarized light microscope; absorption spectrum; and color coordinates (a numerical designation for a particular color determined with the microspectrophotometer). A fiber's dye may be analyzed by thin-layer chromatography, a process by which dye components are percolated through a suitable absorbing medium to separate them.[27]

Whether the task at hand is simply to identify a fiber or whether it is to make a comparison between known and questioned samples, the criminalist seeks to learn the type of fiber, whether natural or man-made; the generic type and subtype; the fiber's color and shade; the expected use or application (i.e., type of textile material); the fiber's manufacturer

FIGURE 3.5. Photomicrograph of cotton fibers (100x), showing their distinctive twist

and the period of manufacture, if determinable; and the relative commonness or rarity of the fiber.[28]

That transfer of fibers occurs is not just a theory (Locard's Exchange Principle) but has been established by scientific tests. After transfer, the persistence of fibers depends on numerous factors. In the case of a criminal wearing clothes with tell-tale fibers, studies show that some 80 percent can be expected to be lost in four hours, with just 5 to 10 percent remaining at the end of twenty-four hours. Hence, it is important to collect evidence from both complainants and suspects as quickly as possible following an alleged offense. Fibers may be retrieved by tape lifts—using clear adhesive tape pressed to the surface of the item, which is searched in grid fashion—or by vacuuming. Brushing and shaking are not recommended. Fibers can be collected on a sheet of clean, prefolded paper that is then placed in an evidence bag or envelope and properly labeled.[29]

The value of fiber evidence will be demonstrated in the case study at the end of this chapter, the Atlanta Child Murders case, in which proper collection and identification of a distinctive series of fibers on the clothing of victims provided considerable information on the killer's environment—long before his arrest provided confirmation of the accuracy of the fiber evidence.

FIGURE 3.6. Photomicrograph of synthetic fibers, semi-dull nylon (100x). Note the delusterant (dulling) pigments embedded in the fibers.

SOIL AND BOTANICALS

Other trace evidence that may be carried to or from a crime scene—particularly on shoes or trousers—includes soil evidence and what is broadly termed botanical evidence or simply "botanicals."

Soil. What is called *soil* is actually a mixture of weathered and decaying rock (including a variety of minerals such as quartz, mica, and feldspar), humus (decomposed organic material), other vegetable matter (bits of leaves, pine needles, plant stems, pollen grains, etc.), and often particles of manufactured materials (asphalt, paint flakes, glass fragments, and so on).[30] The possible constituents are so varied that the forensic definition is quite broad: "Soil may be thought of as including any disintegrated surface material, both natural and artificial, that lies on or near the earth's surface."[31]

Like the analysis of hair and fiber evidence, forensic soil analysis depends on comparing questioned specimens with known samples (soil must be collected in suitable clean containers). A variety of comparative techniques is employed. First, the criminalist considers gross visual appearance, comparing the color and texture of the sample, keeping in mind that moisture darkens the color of soil, making it necessary that all

FIGURE 3.7. Collection of trace evidence such as hair and fibers is typically carried out through use of a vacuum collection system similar to that being used here by a forensic trainee. (Courtesy of Kristin Hayes Wolf, Orange County Sheriff's Office, Orlando, Florida.)

samples be dried under exactly the same lab conditions. Low-power stereomicroscopic examination reveals the various plant, animal, and manmade materials contained in each sample. Technical microscopic examination requires the knowledge of a geologist to use high-powered and polarized microscopy to identify rock and mineral particles. Particle size determination is accomplished by passing samples through a nested series of sieves, with the opening size graduated from largest, on top, to smallest. Instrumental and chemical analyses are conducted, employing such techniques as spectrographic analysis, x-ray diffraction, differential thermal analysis, and both qualitative and quantitative chemical analysis.[32]

X-ray diffraction is an excellent technique for analyzing crystalline materials—such as minerals—by passing x-rays through them and then measuring the angle at which the x-rays were diffracted (or deflected); each crystalline substance yields its own distinctive "diffractogram." Differential thermal analysis depends on a sample's ability to absorb or release heat as it is continuously heated in a specially designed furnace

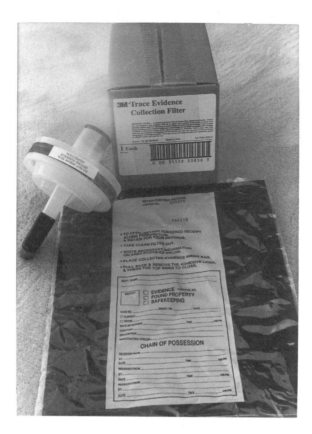

FIGURE 3.8. Vacuum attachments that limit the chance of cross-contamination of sweepings, as well as providing documentation for the chain of custody, are commercially available.

over a wide temperature range; a specimen may become dehydrated or decomposed, undergo transformation of its crystalline structure, melt or boil, or undergo other reactions—at a particular temperature. Some of these reactions are *endothermic* (they *take on* heat) while others are *exothermic* (they *give off* heat). Differential thermal analysis is a useful technique for comparing a variety of materials, including soil and other geologic material, rubber, polymers, ceramics, and many other substances.[33]

Regarding the evidence required to reach a positive conclusion in a soil comparison, one authority states: "Considering the vast variety of minerals and rocks and the possible occurrence of man-made debris that may be present in soil, the forensic geologist is presented with many points of comparison between two or more specimens. The number of comparative points and their frequency of occurrence must all be taken into consideration before similarity between specimens can be concluded and the probability of common origin judged."[34] To the question of whether there is another place on earth where a soil sample could exist that is exactly similar to one in question, the authors of *Forensic Geology* respond: "In most cases, a question of this type can only be answered yes. Soils and related material are not generally individual type items, such as fingerprints, and evidence based on their study can only be stated in terms of samples being similar or dissimilar, comparable or not comparable. Soil samples approach individuality when several methods are used and unusual minerals or combinations of minerals are found."[35]

A rape case in Canada illustrates the value of soil evidence under unusual circumstances. The knees of the suspect's pants were encrusted with soil. Interestingly, the sample of the left knee was significantly different from that of the right knee. At the crime scene, a pair of knee impressions was found, and samples were collected from each. The sample from the left knee impression compared favorably with that from the suspect's left trouser knee, while the sample from the right impression was comparable to that from his right trouser knee, even though the samples from each knee of the trousers differed significantly. This evidence strongly indicated the presence of the suspect at the scene.[36]

Botanicals. As indicated in the discussion on soils, plant matter may be among the disintegrated materials making up soil, or it may be found as trace evidence on its own. The term botanicals refers to anything of the vegetable class—for our purposes, anything such as leaf fragments, pine needles, tree bark, weed stems, seeds or segments of seed pods, bits of moss or lichen, remnants of flower petals, and even pollen. According to *Forensic Geology,* "In forensic studies there is a natural tendency to focus on the mineral portion of the soil, but the character of the soil organic matter should not be discounted. A close relationship exists between soils and vegetation, and from the character of organic matter alone we can make projections as to the possible origin of organic material. The character of the humus layer itself can also be used as an important criterion."[37] Unfortunately, the identification of fragmented vegetable matter typically requires considerable expertise. It may be possible, however, to enlist the aid of a botanist employed at a local arboretum, natural history museum, or university department of botany.

Pollen, produced in the anthers of blooming flowers, is especially useful in determining whether a suspect was in an area where particular flowering plants grow. Pollen grains may be collected by vacuuming the suspect's clothes.[38] Another method is to use clear adhesive tape in a grid pattern as described earlier in the discussion of fibers. The identification of pollens is accomplished by use of the laboratory microscope. Comparison may be made with specimens collected from the scene, from reference slides, or from photomicrographs in published reference sources such as Walter C. McCrone's monumental *Particle Atlas.* Identification may be made as to species. For example, ragweed can be distinguished from corn or oak, but, apart from size of grains, all types of grass pollen look alike.[39]

The investigator should consider whether a plant in question is pollinated by wind (*anemophilous*) or by insect (*entomophilous*), since in the former instance a person may acquire the pollen simply by being in the locale of such plants, whereas in the latter, actual contact is necessary for the pollen to transfer. According to microanalyst Skip Palenik:

> The potential for such information should not be overlooked when trying to corroborate a story or alibi. Experiments have shown that one may walk through a field of flowering dandelions without leaving any, or at most, a few grains of pollen on the clothing. If one kneels momentarily in the same field, however, the knee in contact is covered with thousands of grains, which are difficult to remove completely by brushing. A suspect in a rape, for example, which is known to have occurred in such a field would have difficulty explaining the presence of significant quantities of indicative entomophilous pollen on his upper clothing.[40]

An interesting case in which one of us (J.F.) was involved illustrates that trace material does not always fall into an easily classifiable category; nevertheless, its evidential value is not thereby diminished. The case began on February 2, 1993, when the partially clad, battered body of a woman was discovered at the rear of an unoccupied residence. A concrete block was the apparent murder weapon. Fecal matter from the victim was found on the ground and smeared on an exterior wall of the residence. Investigation revealed that the night before her death the victim had left a local bar in the company of a man with whom she had been drinking. When found, the man told police that he and the woman had gone to the deserted place and sat on the carport, whereupon she had promptly passed out. He decided to walk home, leaving her asleep but alive. The man agreed to let police examine the clothing he had worn but claimed he could not find his shirt; also the jeans he offered were suspiciously clean, inconsistent with his having sat on the carport. He offered two pairs of tennis shoes. One shoe had a slight bloodstain and also had apparent fecal matter embedded in the sole.

DNA tests of the bloodstain indicated the blood was similar to the victim's and occurred in 11 percent of the white population. Although

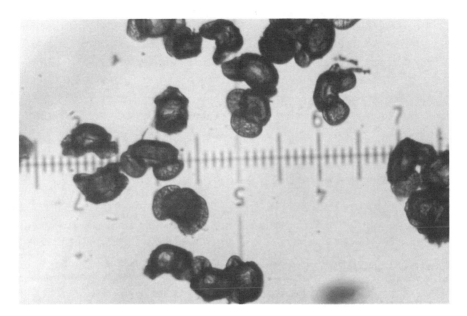

FIGURE 3.9. Photomicrograph of pine pollen (100x)

this was deemed by itself insufficient evidence to make an arrest, there remained for examination the material adhering to the tennis shoe. The shoe and plastic vials containing feces from the scene were sent to a private laboratory, McCrone Associates in Chicago; subsequently, dirt and debris samples from five areas of the crime scene were also sent. First, both sources of feces were positively identified as such by using a test for urobilin (a brownish resinous pigment found in human and animal waste). Because the feces lacked the hairs and undigested food typical of animal feces, the McCrone microscopist, Richard E. Bisbing, concluded it was human in origin. Moreover, both the feces from the scene and that from the suspect's tennis shoe contained certain embedded debris that was typical of that at the crime scene: quartz sand, plant and woody tissue fragments, pollen, trichomes (hairlike outgrowths from the epidermises of plants), and limestone fragments (from gravel), among other materials. Although Bisbing's report mentioned that "none were in particular abundance and none could be associated with the dirt and debris from the scene with any specificity,"[41] the evidence, though far from conclusive, added to that already developed: that the suspect was the last known person to see the victim alive, that his story of leaving was improbable, that it was unlikely that someone else had happened upon the victim and murdered her, that the suspect's clothing was suspiciously missing, that there was a trace of blood consistent with the relatively rare blood of the victim, and finally, that fecal matter with sand and botanicals consistent with that at the scene was found on his shoe, just as it could be expected to be found on the murderer's shoe. Tying this biological, mineral, and vegetable matter to the suspect added to the weight of evidence against him and resulted in his arrest and subsequent guilty plea.

GLASS, PAINT, AND OTHER TRACE EVIDENCE

In addition to the materials discussed thus far, a variety of additional substances—indeed, virtually anything—may appear as trace evidence. **Glass.** Glass may play a role in a variety of crimes: a broken automobile headlight in a hit-and-run case, a smashed basement window in a burglary, or a fractured lamp in a homicide. Glass fragments may be recovered at the scene of the hit-and-run or be collected from the suspect's clothing or shoe soles in the burglary or homicide. In one burglary case, tiny glass fragments were found embedded between the head and handle of a hammer found among a suspect's tools, and these had the same properties as pieces of glass collected at the scene.[42]

Glass fragments—questioned and known specimens—are compared for a common source of origin in a series of scientific tests that determine the chemical and physical properties of each. Microscopical determination may be made of the refractive indices. If both specimens refract (bend) rays of light at the same angle, they may be said to have the same refractive index. The respective densities may be determined. This comparative technique employs a glass tube, closed at one end, that is filled with liquids of different densities, arranged in order of heaviness, with the densest one at the bottom. The two particles to be compared are then introduced into the top of the tube, and if they float at the same level they may be considered to have equal densities.[43]

If the fragments are dissimilar in either refractive index or density, then they can be excluded from having a common source. If they are alike, the refractive index and density values can be correlated to their frequency of distribution so that the likelihood of their having the same source can be considered. The refractive index alone is highly distinctive, as shown by almost twelve hundred glass samples analyzed by the FBI laboratory. (For example, a fragment of glass with a refractive index of 1.5290 was found in just one out of twelve hundred samples, while one with an index of 1.5184—the most common among the glasses—was found in only twenty-eight of the twelve hundred specimens. The refractive index and density taken together would provide an even greater measure of distinctiveness.)[44]

In the case of larger fragments, it may be possible to demonstrate their commonality of origin by fitting the pieces together like a jigsaw puzzle—a technique called *fracture matching*. For instance, two large pieces of an automobile headlight were recovered from the scene of a hit-and-run accident in Kansas City, Missouri. These fit exactly with the remaining pieces that were found in the light housing and grille of the suspect's vehicle.[45]

Viewing glass evidence from an even larger perspective, the investigator is sometimes challenged to determine whether or not a crime has occurred and to assess the credibility of persons making statements at the scene. This situation is illustrated by a case in which a man claimed to have been fired upon three times from outside his residence. He said he had returned one shot at the assailant. The suspect told a similar story, but he claimed to have been shot at three times while walking by the residence and said he had fired one shot in return. The window provided the evidence that resolved the matter. It was possible to tell that one bullet had been fired first because the radial fractures of the first bullet had stopped those fired later. Nothing was known, however, as to when the other two were fired. Nevertheless, it is possible to determine from which side a bullet entered a pane of glass because there is a characteristic cone-shaped depression on the *exit* side; in this case there were three such areas on the inside of the glass indicating that those holes were fired from the outside, and one conical depression on the outside indicating a bullet fired from the inside. Therefore, the glass evidence supported the account of the original complainant.[46]

Paint. As evidence, paint is commonly associated with breaking-and-entering and hit-and-run cases. When it is transferred to an object, oil-based paint tends to produce *smears*, and automobile paint usually shows up as *chips*. A burglar may have paint (or varnish, etc.) traces on his clothing as a result of the force used to pry open a door or open a cabinet or safe. Alternatively, if the tool used has a painted surface, traces of the paint may be left at the point of forced entry. Also, if a painted object is used to assault someone, paint particles may be found on the victim's clothing or in the wounds.[47]

Comparison of paint samples is usually effected in one of three ways: by identification of its chemical and physical properties; by the number of coats and the sequence of colors they represent, seen as layers in a chip; and by fracture matching (i.e., fitting a chip or flake to an area of paint loss).[48]

The first two of these are best accomplished by comparing known and questioned specimens side by side using the stereoscopic microscope. The color, the texture of the surface, and the sequence of color layers are the primary features of interest. Where there is just a single layer, chemical solubility tests may also reveal the type of paint, and a sophisticated technique called pyrolysis gas chromatography can distinguish paint formulations. X-ray spectroscopy is an instrumental method useful for identifying the pigments in paint. It has been estimated that, if all the analyses mentioned in this paragraph are employed and a questioned automobile paint compares favorably to a sample taken from a suspect's car, the odds are about sixteen thousand to one against the questioned paint's coming from another vehicle.[49]

In collecting paint specimens, the adhesive tape method should be avoided because it makes later separation difficult. Scraping off of samples damages the layers and should not be undertaken. Instead, samples should be collected by chipping down to the surface of the underlying wood or metal. Paint samples should be removed with a wooden implement, if possible, to avoid metal traces that would interfere with analyses; if necessary, a steel scalpel can be employed and delivered with the samples. In difficult cases, a small piece of the surface bearing the paint should be chipped or cut off.[50]

A case from Manhattan illustrates the importance of paint evidence. Following the early-morning slaying of three men atop a roof garage, a woman's body was discovered in a distant alley. She had been shot like the men, and it was evident that her body had been removed from the garage and dumped in the alley. A few days later a suspect was arrested in Kentucky, and his black van was processed as a crime scene. Vacuum sweepings from the interior included gray metallic/black paint chips. Trace evidence from the woman's body—including similar paint chips—was also collected, some at autopsy and some from the victim's clothing, processed by a criminalist. Chemical, microscopical, and instrumental tests showed the paint samples to be similar in every respect.

During the trial, in addition to strong circumstantial evidence, additional trace evidence was presented, including clear, amber, and green glass fragments, sawdust, urethane foam, cellophane, blue olefin plastic, brown and white dog hairs, and white seeds. All items were found both in van sweepings and on the body of the victim. In the case of the white seeds, a botanist testified that, although he could not identify them, the one from the victim (discovered in her mouth at autopsy) and those from the van were from the same species, which he thought was probably a rare wildflower. The defendant was convicted and sentenced to a prison term of one hundred years.[51]

Other Man-made Materials. As the foregoing case shows, all types of trace evidence may be found at a crime scene—in that case the interior of a van. (Curiously, no fiber evidence was reported.) Note that in addition to most of the categories of evidence we have discussed in this chapter thus far, there were various types of manufactured materials—cellophane, urethane foam, blue olefin plastic—that also played an important evidentiary role.

The variety of man-made materials is too great to catalogue here, but a short list would include bits or fragments or other traces of materials such as rubber, concrete, plaster, wire or other metal products, numerous plastics including Styrofoam, paper, soap, waxes and polishes, safe insulation, lipstick and other cosmetics, cigarette ash, and so on and on. Some manufactured or *processed* materials are often classified as botanicals; these include wood shavings and sawdust, tobacco, coffee, flour, starch, etc. The possibilities recall the motto of the great German chemist Justus von Liebig (1803-1873): "Dust contains in small all the things that surround us."[52]

CASE STUDY: THE ATLANTA CHILD MURDERS

Over a twenty-two-month period beginning in July 1979, Atlanta, Georgia, was the scene of the murders or disappearances of thirty African American children and young men. The serial killings made national and international headlines. As public pressure to solve the crimes intensified, a special Atlanta Task Force made up of detectives and other investigators from a number of law enforcement agencies was set up to solve the crimes. Even President Ronald Reagan—"hardly the black community's most sensitive friend," as one writer put it—became involved, pledging $1.5 million in federal funds for the investigation.[53] The modus operandi in the murders was not always the same, especially the manner in which the victims were killed, leading detectives to believe early in the investigation that they were looking for multiple suspects.[54]

In time, a discovery by criminalist Larry Peterson set investigators on the track of a single killer. Peterson and his colleagues at the Georgia State Crime Laboratory began to notice on the clothing and bodies of the murder victims a number of distinctive yellowish-green nylon fibers, together with some violet acetate fibers. Although these were not the only similarities among the murders, the fiber evidence was especially demonstrative and held out the promise that the source of the fibers might be discovered and a link proven. By 1981 the criminalists had tentatively concluded that the yellowish-green fibers, because of their coarseness and lobed, cross-sectional appearance, came from a rug or carpet. Yet they could not locate the manufacturer, despite photomicrographs of the fibers being distributed to chemists and other experts in the textile industry.[55]

In February 1981, information about the fibers, which was kept secret so as to not tip off the murderer, was leaked to the press and published in an Atlanta newspaper. Until then, the bodies always had been found fully clothed, but after publication, within a two-and-a-half-month period that saw nine more murders, the nude bodies of seven victims were recovered from Atlanta-area rivers—an obvious attempt to destroy the incriminating evidence.

In retrospect, however, the tip-off proved beneficial because the murderer's response gave police an investigative strategy: They began to stake out bridges from which bodies might in the future be dumped. In the early hours of May 22, 1981, a four-man team of Atlanta police and FBI agents were maintaining surveillance of the James Jackson bridge spanning the Chattahoochee River in the northwestern part of the city. The personnel were located under the bridge and at each end. About 2:00 A.M. the surveillance team heard a loud splash. They stopped the only car that had been on the bridge at that time, driven by one Wayne Bertram Williams, a local music promoter. Police said Williams told them he had dumped garbage off the bridge, but he later denied making the statement. Two days later, police recovered the body of a man later identified as Nathaniel Cater. In his hair they discovered a yellowish-green nylon fiber.

Wayne Williams soon became aware that investigators were following him, and he led them on wild-goose chases around Atlanta. Williams was an avid photographer, with a darkroom in his home, and before police could obtain a warrant to search his premises, he was seen burning a quantity of photos in his backyard. He also washed out his station wagon. He agreed to take a polygraph (or so-called "lie detector") test, which yielded inconclusive results. (Later police would find among his possessions books that explained how to beat a polygraph.)[56]

As part of the circus atmosphere that prevailed at the time along with Williams's bold, defiant antics, "psychics were swarming around . . . many dramatically contradicting each other."[57] The clairvoyants were of no help to the police and were, in fact, a nuisance. That includes the self-styled "police psychic," Dorothy Allison.[58] On June 3, 1981, after obtaining a warrant, police searched Williams's home, where he lived with his parents, and also his automobile. Over time, several more searches would be made, and fibers by the hundreds of thousands were eventually collected by criminalists.

On the evening after the first search, Larry Peterson was working late on the fiber evidence when he thought he had made some successful matches between questioned fibers from victims and known fibers from the Williams residence. Those included the yellowish-green carpet fibers, fibers from the bedspread on Wayne Williams's bed, and hairs from the Williams family's dog. Despite the late hour, Peterson phoned Special Agent Harold A. Deadman from the FBI's Microscopic Analysis Unit, who was in Atlanta working on the Williams case.[59] "Peterson called me at about one-thirty in the morning," Hal Deadman recalled, "and told me very matter-of-factly, 'I've made some matches. You'd better come over here.' I went right over to the lab and looked at them. They looked fine. At that point both Larry and I were convinced that someone in the Williams environment was involved in the murders. The chance of this being just a coincidence was essentially zero."[60]

Soon the source of the yellowish-green fibers was identified. The Wellman Corporation, a Massachusetts firm, had originally made the nylon product, known as 181B. Wellman then sold this in the form of yarn to West Point Pepperell, a carpet company in Dalton, Georgia. The carpet company in turn used the yarn in its Luxaire line, producing an "English Olive" Luxaire carpet that was sold in limited numbers in the Atlanta area.[61]

To determine just how unusual the Williams's carpet was, prosecutors attempted to estimate the numerical probability—"something never before done in connection with textile materials used as evidence in a criminal trial."[62] It was found that West Point Pepperell had sold retailers in ten southeastern states 16,397 square yards of English Olive carpet (containing the Wellman 181B fiber) in 1971 and 1972. Assuming an equal division of sales among states and that installation was for one twelve- by fifteen-foot room in each instance, about eighty-two rooms could be expected to have this carpet in Georgia. Since statistics showed 638,995 occupied housing units in the metropolitan Atlanta area, the odds were just 1 in 7,792 that a similar carpet would be found by random selection. (Actually the odds were greater, because some residences might have used more than the estimated amount, thus meaning there would be fewer residences in the computation, and no doubt some of the carpet would have been discarded in the intervening eleven years.) Moreover, these odds would be multiplied by each type of hair and fiber discovered on the victims that could be linked to Wayne Williams.

In all, twenty-eight different fibers, plus dog hairs, found on one or more victims were identified as consistent with nineteen objects located in Williams's home and automobile.[63] "On some of the victims we found as many as ten fibers that could be directly related to Williams's home or a car he had in his possession at the time of that murder," Deadman said. "The chances that those ten fibers could have come from a place other than Williams's home or car are just about impossible."[64] Deadman also points out:

> The combination of more than 28 different fiber types would not be considered so significant if they were primarily common fiber types. In fact there is only 1 light green cotton fiber of the 28 that might be considered common. This cotton fiber was blended with acetate fibers in Williams' bedspread. Light green cotton fibers removed from many victims were not considered or compared unless they were physically intermingled with violet acetate fibers which were consistent with originating from the bedspread. It should be noted that a combination of cotton and acetate fibers blended together in a single textile material, as in the bedspread, is in itself uncommon.[65]

The trial of Wayne Williams focused on just two victims who were included in the July 1981 indictment against him: Jimmy Ray Payne, age 21, and Nathaniel Cater, age 28. The prosecution built its case against Williams in the death of these men by "educating" the jury regarding textile fibers. This involved having representatives from Wellman and West Point Pepperell Corporations testify about their products and having forensic experts explain how fibers are analyzed in the crime laboratory. They calculated not only the individual probabilities (about 1 in 7,792 in the case of the yellowish-green carpet fiber, and 1 in 3,828 in the case of carpet fibers from Wayne Williams's 1970 Chevrolet station wagon), but they also explained how, if different fiber types from more than a single object are found and each links the victim to a particular environment, the strength of the evidence is multiplied.[66] In other words, "as the number of different objects increases, the strength of an association increases dramatically. That is, the chance of randomly finding several particular fiber types in a certain location is much smaller than the chance of finding one particular fiber type."[67] Put still another way, only a limited number of people would have the English Olive carpet; of those, how many could also be expected to have the violet-green bedspread? And of those few who might have the same carpet and bedspread, how many would also drive a 1970 Chevrolet station wagon? And of those, how many would own the numerous other items plus a German shepherd? And so on.

According to one account of the trial: "In addition to being used to link Williams to the victims, fibers were also used to establish a time frame. Nine victims were linked by automobile carpet fibers to three different cars used by Williams. When Williams was using a rented car, fibers that could be matched to carpeting from that car were found on victims. A noose comprises millions of fibers woven together to make an extremely strong rope. The hair and fiber evidence was forming a noose that was tightening around Williams's neck."[68]

Following the testimony regarding Payne and Cater, the prosecution asked permission from the court to introduce evidence of ten other victims whose cases were similar to those at trial. Georgia law permits the evidence of another crime to be introduced "if some logical connection can be shown between the two from which it can be said that proof of the one tends to establish the other as relevant to some fact other than general bad character."[69] The court permitted introduction of the other evidence in an attempt to establish a "pattern" or "scheme" of murder that included the two homicides for which Williams was being tried. Each of the victims' bodies had fibers that could be linked to Williams. Another pattern was that most of the twelve victims were either strangled or asphyxiated and their bodies disposed of in remote areas.[70] To make their case, the prosecution utilized over forty charts and over three hundred fifty photographs or photomicrographs to illustrate to jurors what the criminalists could see through their microscopes in the laboratory. Table 3.1 shows a partial list of possible fiber sources and indicates on which of the twelve victims mentioned in court they were found.[71]

The defense sought to discredit the fiber evidence. Whereas the prosecution asked the jury to *multiply* (one fiber times another), the defense team sought to *divide* and thus conquer. They argued that many people might have this or that fiber in their homes or vehicles. They suggested that Payne's and Cater's bodies had another possible source for multiple and varied fibers: the river! The prosecution countered that argument by citing other victims whose bodies were *not* dumped in the river but rather in wooded areas or along streets yet who had the same fibers.[72]

TABLE 3.1.
The Prosecution's Fiber Links

Possible Fiber Sources	VICTIMS											
	1	2	3	4	5	6	7	8	9	10	11	12
Violet and green bedspread	x	x	x	x	x	x	x	x	x	x	x	x
Green bedroom carpet	x	x	x	x	x		x	x	x	x		x
Pale green carpet squares in office/workroom area	x							x		x		
Yellow blanket found under bed	x	x	x	x	x		x					
Trunk liner from 1979 Ford LTD										x	x	
Red carpet from 1979 Ford LTD											x	
Dark carpet from 1970 Chevy station wagon		x	x	x			x					
Yellow toilet cover					x					x	x	
Gray glove found in station wagon		x					x					
Brown waist and collar from leather jacket							x					
Yellow kitchen rug									x			
Blue acrylic throw rug			x									
Trunk liner from 1973 Plymouth Fury												x
Fibers found in debris from station wagon (several colors)		x	x	x	x	x	x	x		x		
White polyester fibers found in debris from rug in station wagon cargo area							x	x		x		
Animal hairs consistent with suspect's dog	x	x	x	x	x		x	x	x	x	x	x
Bedspread hanging in carport area				x	x							

Source: Hyde Post and David B. Hilder, "Fibers Found on Victims Form Links in Williams Case," *The Atlanta Journal*, February 2, 1982, p.1A, with modification.

Nor was the fiber evidence all that the prosecution had against Williams. Williams fit the behavioral profile of the killer in various ways, including racially. Though the public perception was that the killing of black children and men was the work of a white supremacist, the FBI Academy's Behavioral Sciences Unit showed that such murderers usually attack victims from their own racial group.[73] Witnesses testified that they had seen Williams with some of the victims, and, of course, there was the fact that after his arrest the killings had stopped.[74] Williams was put on the stand and made a good witness until, during cross-examination, he lost his temper and began ranting. He looked like someone out of control, and he lost much of the goodwill he had built with the jury.[75] The jury deliberated eleven hours before finding Wayne Williams guilty of both murders. He was sentenced to two life terms, which he is serving consecutively at the Valdosta Correctional Institution in southern Georgia. Although, like so many criminals, he still maintains his innocence,[76] the fiber evidence demonstrates otherwise. According to one writer: "The knowledge that police had found a few fibers on his victims had forced Wayne Williams to change his killing patterns, and caused him to be caught on that bridge above the Chattahoochee River. And it was those small fibers, fibers no different from those you can pick off your clothing right now, that resulted in Williams being convicted of murder and sentenced to spend the rest of his life in prison."[77]

Apart from the conviction of Williams, and with it the end of a series of slayings that outraged the world, the case had another important outcome. "Perhaps in no other major case," according to one forensic text, was the collection and analysis of fiber evidence "as critical to a successful prosecution" as it was in the Wayne Williams trial.[78] Although the case has remained controversial,[79] it set new precedents for the presentation of fiber evidence—indeed of trace evidence in general—that will no doubt help bring additional killers and other criminals to justice in the future.

On July 10, 1998, a Georgia circuit judge upheld the 16-year-old convictions of Wayne Williams, whose lawyers had argued that prosecutors had withheld evidence in the case against him. Judge Hal Craig noted that "nothing has weakened the testimony about the scratches down petitioner's arms, consistent with defensive marks left by strangling victims." He termed the fiber evidence "the strongest scientific link in this case."[80]

NOTES

1. Gaylord Johnson and Maurice Bleifeld, *Hunting with the Microscope*, rev. ed. (New York: Arco, 1978), 5-11.

2. *Lincoln Library of Essential Information* (New York: Frontier Press, 1946), s.v. "Microscope."

3. Johnson and Bleifeld, *Hunting with the Microscope*, 8-11; Peter R. DeForest, "Foundations of Forensic Microscopy," in *Forensic Science Handbook*, ed. Richard Saferstein (New York: Prentice-Hall, 1982), 423-24, 429.

4. Joe Nickell, *Detecting Forgery* (Lexington: Univ. Press of Kentucky, 1996), 145; *1996 Optics & Optical Components Catalog* (Barrington, N.J.: Edmund Scientific Co.), 145.

5. Nickell, *Detecting Forgery*, 151; DeForest, "Foundations," 434.

6. DeForest, "Foundations," 434.

7. Raymond C. Murray and John C.F. Tedrow, *Forensic Geology* (New York: Prentice-Hall, 1992), 109-13; Richard Saferstein, *Criminalistics: An Introduction to Forensic Science,* 5th ed. (Englewood Cliffs, N.J.: Prentice Hall, 1995), 182-83; William E. Ford, ed., *A Textbook of Mineralogy* by Edward S. Dana, 4th ed. (New York: John Wiley & Sons, 1932), 250-63.

8. DeForest, "Foundations," 433-34; Saferstein, *Criminalistics,* 180-82.

9. DeForest, "Foundations," 434.

10. Ibid., 433; Nickell, *Detecting Forgery,* 145-46.

11. DeForest, "Foundations," 436; Saferstein, *Criminalistics,* 5.

12. Lincoln Library, s.v. "Microscope"; Murray and Tedrow, *Forensic Geology,* 119-21; Philip Paul, *Murder under the Microscope: The Story of Scotland Yard's Forensic Science Laboratory* (London: Futura, 1990), 120.

13. Murray and Tedrow, *Forensic Geology,* 120-21.

14. Saferstein, *Criminalistics,* 183-85; quotation, 184.

15. Skip Palenik, "Microscopy and Microchemistry of Physical Evidence," in Saferstein, *Forensic Science Handbook,* 176-83.

16. Ibid., 205-6.

17. Richard E. Bisbing, "The Forensic Identification and Association of Human Hair," in Saferstein, *Forensic Science Handbook,* vol. 2, 188.

18. Charles R. Swanson Jr., Neil C. Chamelin, and Leonard Territo, *Criminal Investigation,* 4th ed. (New York: McGraw-Hill, 1988), 92.

19. Bisbing, "Forensic Identification," 203.

20. Ibid.

21. Charles E. O'Hara, *Fundamentals of Criminal Investigation,* 3d ed. (Springfield, Ill.: Charles C. Thomas, 1973), 757.

22. Swanson, Chamelin, and Territo, *Criminal Investigation,* 92-93; Saferstein, *Criminalistics,* 212.

23. Swanson, Chamelin, and Territo, *Criminal Investigation,* 67.

24. For a more complete classification chart, see James Robertson, ed., *Forensic Examination of Fibres* (New York: Ellis Horwood, 1992), 1-40; chart, 2.

25. Saferstein, *Criminalistics,* 213. Plant fibers, notably cotton and linen, are also found in rag paper and so are also of interest to the questioned-document examiner. See Nickell, *Detecting Forgery,* 151-52, 178-79.

26. Barry D. Gaudette, "The Forensic Aspects of Textile Fiber Examination," in Saferstein, *Forensic Science Handbook,* 222-39.

27. Ibid., 221-54.

28. Ibid., 221; Robertson, *Forensic,* 100-101.

29. Robertson, *Forensic,* 44-52.

30. Barry A.J. Fisher, *Techniques of Crime Scene Investigation,* 5th ed. (New York; Elsevier, 1992), 190; Saferstein, *Criminalistics,* 112-13.

31. Saferstein, *Criminalistics*, 112.

32. Ibid., 113-14; Fisher, *Techniques*, 190; Fred E. Inbau, Andre A. Moenssens, and Louis R. Vitullo, *Scientific Police Investigation* (New York: Chilton, 1972), 110.

33. Murray and Tedrow, *Forensic Geology*, 123-28.

34. Saferstein, *Criminalistics*, 114.

35. Murray and Tedrow, *Forensic Geology*, 161-62.

36. Ibid., 23

37. Ibid., 65.

38. Fisher, *Techniques*, 194-95.

39. Murray and Tedrow, *Forensic Geology*, 68.

40. Skip Palenik, "Microscopy and Microchemistry of Physical Evidence," 191.

41. Richard E. Bisbing, report to Detective John Linnert, Orange County Sheriff's Office, Orlando, Florida, March 1, 1994; case-outline transmittal report by Detective John H. Linnert, April 14, 1994.

42. Inbau, Moenssens, and Vitullo, *Scientific Police Investigation*, 102-3.

43. Ibid., 103-4.

44. Saferstein, *Criminalistics*, 103-5, graph 107.

45. Swanson, Chamelin, and Territo, *Criminal Investigation*, 64-66; illus., 65.

46. Ibid., 66-67; our illustration is adapted from theirs.

47. Ibid., 63; Inbau, Moenssens, and Vitullo, *Scientific Police Investigation*, 105.

48. Saferstein, *Criminalistics*, 229-35; illus., 230 (showing a fracture match).

49. Ibid., 229-35.

50. Swanson, Chamelin, and Territo, *Criminal Investigation*, 63-64.

51. N. Petraco, "Trace Evidence—The Invisible Witness," *Journal of Forensic Sciences*, vol. 31 (1986): 321; reprinted in Saferstein, *Criminalistics*, 235-38.

52. Justus von Liebig, quoted in Jurgen Thorwald, *Crime and Science: The New Frontier in Criminology* (New York: Harcourt, Brace & World, 1966), 285.

53. Martin Fido, *The Chronicle of Crime* (New York: Carroll & Graf, 1993), 283.

54. David Fisher, *Hard Evidence* (New York: Dell, 1995), 142.

55. Ibid., 142-44; Harold A. Deadman, "Fiber Evidence and the Wayne Williams Trial," in Saferstein, *Criminalistics*, 74-76.

56. Fisher, *Hard Evidence*, 143; Deadman, "Fiber Evidence," 76; John Douglas and Mark Olshaker, *Mind Hunter: Inside the FBI's Elite Serial Crimes Unit* (New York: Scribner, 1995), 212-13.

57. Douglas and Olshaker, *Mind Hunter*, 211.

58. For a discussion of psychics (including Allison) and the techniques they employ to convince people of their clairvoyance, see Joe Nickell, ed., *Psychic Sleuths: ESP and Sensational Cases* (Amherst, N.Y.: Prometheus Books, 1994).

59. Deadman, "Fiber Evidence," 76; Fisher, *Hard Evidence*, 143-44.

60. Deadman, quoted in Fisher, *Hard Evidence*.

61. Fisher, *Hard Evidence*, 143.

62. Deadman, "Fiber Evidence," 78.

63. Ibid., 78-80, 87; Fisher, *Hard Evidence*, 145.

64. Deadman, quoted in Fisher, *Hard Evidence*, 145.

65. Deadman, "Fiber Evidence," 87.

66. Ibid., 81-82

67. Ibid., 82.

68. Fisher, *Hard Evidence*, 145-46.

69. *Encyclopedia of Georgia Law* (1979), cited in Deadman, "Fiber Evidence," 84.

70. Deadman, "Fiber Evidence," 84-85.

71. Information for table 3.1 from Hyde Post and David B. Hilder, "Fibers Found on Victims Form Links in Williams Case," *Atlanta Journal*, Feb. 2, 1982, 1A, reproduced in Swanson, Chamelin, and Territo, *Criminal Investigation*, 69.

72. Deadman, "Fiber Evidence, 88.

73. Douglas and Olshaker, *Mind Hunter*, 202-4.

74. Fido, *Chronicle of Crime*, 283.

75. Douglas and Olshaker, *Mind Hunter*, 221-22.

76. Ibid., 222.

77. Fisher, *Hard Evidence*, 146.

78. Swanson, Chamelin, and Territo, *Criminal Investigation*, 69.

79. Charisse Jones, "Guilt of Atlanta 'Child Killer' is in Doubt after 15 Years," *USA Today*, December 8, 1997.

80. Rick Bragg, "Convictions Are Upheld in Murders in Atlanta," *New York Times*, July 11, 1998.

RECOMMENDED READING

McCrone, Walter C. *The Particle Atlas.* Vols. 1-5. Ann Arbor, Mich.: Ann Arbor Science Publishers, 1973-1979. A compilation of photomicrographs of various substances to aid identification; a major reference work.

Murray, Raymond C., and John C.F. Tedrow, *Forensic Geology.* Englewood Cliffs, N.J.: Prentice Hall, 1992. A manual for the forensic identification of earth material; includes glossary.

Paul, Philip. *Murder under the Microscope: The Story of Scotland Yard's Forensic Science Laboratory.* London: Futura, 1990. A popular discussion of laboratory research and its application to crime solving.

Robertson, James, ed. *Forensic Examination of Fibres.* New York: Ellis Horwood, 1992. An authoritative text on all aspects of fibers as related to criminalistics.

Saferstein, Richard, ed. *Forensic Science Handbook.* Englewood Cliffs, N.J.: Prentice-Hall, 1982; vol. 2, 1988. A comprehensive text written by noted experts in the various fields of forensic science, including microscopy; paint, glass, soil, hair, and fibers; firearms identification; and other areas of specialty.

4 | FIREARMS

Murders by firearms became common in the seventeenth century after various wars placed such weapons in the hands of the peasant class. Reluctant to return to their life of poverty, many former soldiers became highwaymen—an occupation they invented. Many victims died before anyone conceived of trying to match a gunshot to the person who fired it. Possibly the first recorded instance of this occurred in 1794 in county Lancashire, England. In conducting a postmortem examination of the victim's body, the surgeon discovered in the wound a wad of paper. In those days firearms were of the single-shot, flintlock variety, loading of which involved pouring gunpowder down the barrel, then dropping in the round lead bullet, followed by a wad of paper that was tamped down with a ramrod to pack the charge tightly. In this case the paper wadding had been driven into the wound along with the bullet (no doubt indicating a relatively close shot). When the paper was opened, it was found to be a piece from a street ballad. When a suspect was arrested in the case, authorities found in his coat the remainder of the ballad, which the piece from the wound matched exactly. This convinced the jury of his guilt, and the court sentenced him to death.[1] So began the important branch of criminalistics concerned with forensic firearms evidence.

BASIC ARMS AND AMMUNITION

Firearms and the ammunition they use are of sufficient variety and complexity to require a brief discussion, complete with the basic nomenclature.

Firearms. Firearms may be divided into two main categories: handguns and shoulder firearms.[2] As their name implies, *handguns*, also called pistols, are designed to be held in the hand. The simplest type is *the single-shot pistol*—that is, one that must be loaded and cocked for each firing.[3] Antique flintlock and percussion pistols are of this type. (In percussion, or "cap-and-ball" firearms, a pull of the trigger and the resulting fall of the hammer discharge the explosive cap, which in turn ignites the powder. The percussion cap replaced the troublesome and somewhat unpredictable combination of flintlock and flash pan—the former showering sparks into a small amount of priming powder.)[4] The .44-caliber Derringer pistol that John Wilkes Booth used to assassinate President Lincoln was of this single-shot percussion type (firing a ball that was homemade of Britannia metal, an alloy of tin).[5] Other types of single-shot pistols include certain models used for training or target shooting,[6] typical "zip guns" (homemade pistols),[7] a clandestine "fountain-pen gun," and the notorious sawed-off shotgun.[8]

The great majority of pistols today are multiple-shot weapons, either revolvers or self-loading pistols. *Revolvers* take their name from their revolving cylinder, which has multiple chambers to hold the cartridges. Most revolvers are five- or six-shot. With a *single-action* revolver, the cylinder turns to the next chamber each time the gun's hammer is cocked by the thumb. A *double-action* model does not require manual cocking; each time the trigger is pulled, the hammer is cocked (and, simultaneously, the cylinder revolves to the next position), and then it is released, firing the cartridge. (Double-action revolvers may also be used in a single-action fashion.) To load a revolver, the cylinder is opened, in most models either by pivoting the barrel downward (and with it the cylinder) or by swinging out the cylinder. These are known, respectively, as "top-loading" and "swing-out loading" designs.[9]

A self-loading pistol, also known as "autoloading" or (erroneously) "automatic," takes cartridges from a magazine, sometimes (erroneously) called a "clip," which is located in the handle of the pistol. With each firing, the pistol's own recoil energy moves the slide rearward, ejects the expended shell case, recocks the firing mechanism, and loads a new round into the firing chamber. This type of pistol may be fired as fast as the trigger is pulled, and is often termed a semiautomatic. A truly *automatic* pistol fires continuously while the trigger is held down.[10]

Unlike handguns, *shoulder firearms* have long barrels and are meant for use by both hands. The main types are rifles and smoothbore arms. Rifles take their name from their rifled barrels. *Rifling* is a set of spiral grooves cut into the interior surface of the barrel; the raised areas be-tween the grooves are called *lands*. The latter bite into a bullet as it passes through the barrel and impart a spin or rotation to the bullet that results in a more stable trajectory.[11]

Rifles may be *single-shot* models (mostly small-caliber sporting rifles used by novices); *lever-* or *bolt-action magazine repeaters*, which rely on the manual handling of a lever or turnbolt to extract and usually also eject the fired shell case, cock the firing mechanism, and move a new round into the chamber; or *semiautomatic* and *automatic* rifles, which work like their handgun counterparts. *Assault rifles*, such as the Soviet AK-47 and the American M-16, may be fired either in semiautomatic or automatic mode. *Machine guns* are fully automatic firearms that load their ammuni-tion either from large magazines or belts, and which, due to their heavy recoil, are fired from a tripod or other mounting. In contrast, submachine guns are lighter and are intended to be hand held; they also fire pistol rather than rifle cartridges and therefore have neither the range nor the accuracy of automatic rifles.

The *caliber* of a firearm originally referred to the diameter of the bar-rel bore. Now, because of minor changes in standards over time, it is only an approximation that in English-speaking countries is usually measured in inches (for example, .38-caliber is 38/100 inches); in countries using the metric system, the caliber is expressed in millimeters.[12]

Unlike rifles, *shotguns* have no rifling grooves and so are a type of *smoothbore*, shoulder-fired weapon. They are measured in gauges; the smaller the number, the larger the bore diameter. They may fire round balls or rifled bullets, or, much more commonly, scatter shot. Shotguns may be of the single- or double-barreled type that break down for reload-ing; alternatively, they are repeating models that feed the shells from the magazine by pump action (or, much less commonly, with lever or bolt action) or that operate semiautomatically (as with other semiautomatic weapons).[13]

Ammunition. A cartridge consists of a *case* made of brass (sometimes nickel-plated, to help prevent corrosion), filled with a *propellant* (black powder or smokeless powder), and fitted at the open end with a *bullet* (the projectile); the *primer* is a small explosive charge that sets off the propellant when it is struck by the firing pin. The primer may be located in an area around the outside circumference of the base, as in .22-caliber cartridges, in which case it is called *rimfire* ammunition. Almost all larger caliber ammunition is *centerfire* type—that is, the primer is lo-cated in the center of the base.[14]

Bullets are usually made of lead or a lead alloy and may be *jacketed* (encased with a harder metal to help keep the bullet intact when it strikes a target) or semijacketed (encased only on the sides of the bullet, or, alternatively, only at the nose). *Soft-point* and *hollow-point* bullets are semijacketed varieties intended to mushroom on impact. Other types of bullets may occasionally be encountered.[15]

Cartridge cases may be of three basic shapes: straight (straight-sided in profile; cylindrical), tapered (narrowing toward the top), or bottlenecked (having a sloping shoulder area that reduces the case to a narrow diameter neck). Straight cases are used for all .22-caliber rimfire cartridges as well as most handgun ammunition. Bottleneck cases are usually used for centerfire rifle cartridges because that form provides a much greater ratio of propellant to bullet caliber, adding higher power. Tapered cartridge cases are essentially obsolete.[16] Shotgun ammunition is different from that of other firearms in several respects. Although formerly having full brass cases, shotshells are now made with brass bases to contain the primer, and the remainder is made of heavy paper or plastic. A plain ball or rifled bullet may be inserted in the end of the case, with the bullet having fins or vanes that promote stability. Usually shotshells contain a quantity of *shot,* or round pellets, in sizes ranging from .05 inch for the number 12 to .33 inch for the number 00. The shot is separated by *wads* (discs of felt, paper, or the like) placed between the shot and the propellant and also over the shot to hold the charge in place and provide a proper seal.[17]

Obviously, this discussion of firearms and ammunition is only introductory. There are other types of each, as well as variations on the components, and different systems of classification and nomenclature. The information here, however, may be sufficient for the novice and is intended to provide the interested reader with a sound basis for more detailed study.

BALLISTICS

Murder She Wrote is not the only television program that gets it wrong, but it is a major offender: "You might want to send that pistol to Ballistics, Lieutenant," nudges Jessica Fletcher. "Right. Sergeant, let me know when the Ballistics report is available," barks the homicide investigator. Alas, they don't mean "Ballistics" at all; they mean "Firearms Identification" or "Examination" or "Comparison." Even the intrepid Columbo has misspoken on this important subject. According to an authoritative source:

FIGURE 4.1. A typical firearms section of a crime laboratory, shown with its reference collection of firearms and other essential equipment, such as microscopes and balances. (Courtesy of Dr. John Kokanovich, Mesa Police Department, Mesa, Arizona.)

> For a number of years this branch of investigation was referred to as forensic ballistics, but the use of this terminology should be discouraged. Ballistics is the science of the motion of projectiles, and one versed in this field is both a physicist and a mathematician. Since a competent firearms identification man need not necessarily have either of these qualifications, it is wise to refrain from claiming to be any type of ballistician.
>
> The term forensic was intended to indicate that the expert's ballistic knowledge was limited to the legal field. To avoid misrepresentation it is wise to refer to this science as *Firearms Identification* and to an individual working in this field as a *Firearms Examiner* or *Firearms Identification Technician*.[18]

In other words, retrieving bullets and shell cases from a crime scene and comparing them with known specimens fired from a suspect's pistol have nothing to do with ballistics but everything to do with forensic techniques of examining, comparing, and (potentially) identifying firearms evidence.

There *are*, however, ballistics matters of forensic relevance that are worthy of discussion. First, it is important to understand the main types of ballistics: internal, external, and terminal.[19]

Internal ballistics deals with, first, the forces that set the bullet in motion, helping it to overcome inertia (the tendency of a body at rest to remain so or of a body in motion to keep moving) and, second, the movement of the bullet through the barrel of the firearm. According to one forensic text: "In the broad sense of the word, interior ballistics also includes those actions which take place before the powder charge is ignited. These include the pulling of the trigger, the fall of the hammer and the explosion of the priming mixture."[20] These matters are important to the firearms examiner because, during this interior ballistics phase, markings are put on the shell case and bullet that will be the focus of any effort to match them with others.

Exterior ballistics deals with the bullet's trajectory—its flight from the time it leaves the barrel until it impacts the target. Once again, inertia is involved because there is no driving force acting upon the bullet after it has left the barrel. Trajectory also involves the additional forces of *gravity*, which begins a downward pull on the bullet as soon as it leaves the barrel; air resistance, which causes a slowing of any projectile; *velocity*, the speed of any portion of the bullet's flight (the first being *muzzle velocity*); and *yaw*, a tendency for the bullet to wobble as it first leaves the barrel, usually corrected by the spin imparted to it.

Terminal ballistics is concerned with how the projectile (including shot, wad, etc.) interacts with its target. Relevant forces include *striking energy*, the force exerted as a bullet hits the target; *penetration*, the depth to which the projectile traverses the target material; and *ricochet*, a bullet's glancing off of an object that it strikes at an extreme angle.[21]

Ricochet "is, to the firearms examiner," according to one text, "a very important ballistic factor."[22] Consider this British case, involving the question of ricochet:

> A man discharged a service revolver in a crowd injuring one of the group. He claimed to have fired the gun toward the floor, and if this was true, he was guilty only of the illegal discharge of firearms. If, however, he had fired toward the crowd, he was guilty of a crime punishable by several years imprisonment. The investigating officer found that the bullet had ricocheted off a piece of metal trim near the floor, then off a side wall, the ceiling and a table leg before causing injury. The shooter was guilty of the lesser crime, but only careful investigation revealed what had happened. The firearms examiner should always investigate a crime scene for the possibility of bullets having ricocheted.[23]

Another interesting case of terminal ballistics was one the authors investigated in 1985. We whimsically called it "The Case of the Shrinking

Bullet." It concerned a man who had been shot through the head; the bullet entered the right eye and exited at the back of the skull. The large exit wound was consistent with a bullet having been fired from a .38-caliber pistol found beside the victim's body. However, the medical examiner observed that the entrance wound was only about .25 inches in diameter, too small to have been caused by the revolver in question, but compatible with a .25-caliber weapon.

We reviewed the case for the police department that had jurisdiction. We noted that the entrance wound in the eye socket of the skull was a type known because of its shape as a "keyhole" defect. Research turned up a case in Boston in which a .32-caliber semiautomatic weapon had produced a classic keyhole defect, the circular portion of which measured only one-quarter inch across. The Boston medical examiner, Douglas S. Dixon, verified the facts in that case, although he could not explain the discrepancy—how such a large bullet could leave such a small hole—except by the thin bone possibly shaving off part of the bullet.

We conducted experiments at length using deer skulls and a duplicate .38 revolver. Eventually one of us (J.F.) succeeded in producing a keyhole defect that measured only about .287 inch across. The bullet was recovered intact. In a subsequent article in *Identification News*[24] and in a chapter in a book,[25] we gave this possible explanation for a "keyhole" defect being smaller than the bullet that produced it: "as it strikes the bone tangentially, the bullet's rounded nose creates the initial small hole, and a triangular portion of bone is knocked free (or shattered into fragments) ahead of it. As the main portion of the bullet passes through the now-semicircular opening, it forces open the 'jaws' of the same, thus allowing the larger bullet to pass on through."[26] Such cases illustrate not only the terminal ballistics and related matters of physics involved, but they also demonstrate the need to keep an open mind and to conduct research and even experiments where warranted.

IDENTIFYING CHARACTERISTICS

In contrast to ballistics, the science of firearms examination is concerned primarily with issues of identification. In 1835 the first successful attempt was made to link a bullet recovered from a murder victim's body to a suspect in the crime. (The 1794 case cited at the beginning of this chapter involved matching the paper *wadding* rather than the actual bullet.) The investigator, Henry Goddard, was one of London's Bow Street Runners, a quasi-detective force that served warrants and performed other duties for Bow Street's magistrates (or justices of the peace). Looking at the bullet removed from the body of a murder victim, Goddard noticed a distinctive flaw, a ridgelike blemish. At the home of a suspect, Goddard discovered a bullet mold that had a correspondingly distinctive gouge at the same location as the bullet. When confronted with this powerful evidence, the suspect confessed to the murder.[27]

In 1889 a French professor named Alexandre Lacassagne sought to identify a murder bullet that had been fired through a rifled barrel and had seven resulting longitudinal grooves. Professor Lacassagne was shown revolvers from several suspects, and he picked out one with seven rifling grooves in the barrel. On this evidence, the suspect who owned the revolver was convicted of murder. Today, however, this would be recognized merely as *class* evidence, not *individual* evidence that would constitute full proof.[28]

The case in which this crucial distinction was made took place in 1898. A German chemist, Paul Jeserich, received a bullet recovered from the body of a man who had been murdered near Berlin. In an approach that anticipated modern procedures, Jeserich fired a test bullet into soft material and then, noticing similarities between it and the questioned bullet, compared the two under a microscope. On the basis of agreement between their markings, he testified that the fatal bullet was fired by the defendant's revolver.[29] As one textbook on criminal investigation laments, "Unknowingly at the doorstep of scientific greatness, Jeserich did not pursue this discovery any further, choosing instead to return to his other interests."[30]

Following Jeserich, in May 1912 in Paris, Victor Balthazard lectured at the Congress of Legal Medicine about a case in which he had testified. The professor used enlarged photographs to illustrate some eighty-five points of correspondence between the fatal bullets recovered in the case and test bullets fired from the defendant's revolver. The evidence resulted in a conviction. Balthazard's lecture was instrumental in "ballistics" being accepted as a legitimate branch of forensic science.[31]

In chapter 1 we saw that Calvin Goddard (who appears to have been unrelated to Henry Goddard, the Bow Street Runner) was a pioneer in refining the techniques of firearms identification, particularly in employing the comparison microscope for that purpose. We will look at Calvin Goddard's contributions in more depth in the case study at the end of this chapter.

Bullet Comparison. As a bullet is propelled from a rifled barrel—that is, one having the spiraling set of grooves and lands that constitute rifling—

a series of parallel markings are imparted to it. These are the *land impressions* and *groove impressions* that constitute class characteristics (which will be similar to those on other firearms of the same make and model) and also certain parallel scratches. These *striae, or striations*, as they are known, are the result of imperfections in the bore—tiny flaws in the lands and grooves that are placed in the barrel both during manufacture and through subsequent use. These striations represent individual characteristics that make it possible to match a test bullet with a questioned one and demonstrate that they came from the same firearm to the exclusion of all others.[32] The class characteristics are useful in identifying the type of weapon. The number of lands and grooves in firearms, which vary from two to twenty-two, can help in identifying the make and model, as can the widths of these impressions. The direction of spiral of the rifling is also important. Rifling with a left-hand twist is often called "Colt-type rifling," while that with a right-hand spiral is often referred to as "Smith and Wesson–type rifling." The degree of twist occasionally may be useful as well.[33]

After the firearms expert has determined that the class characteristics of the markings on the questioned bullet are compatible with those of the test bullet (which is fired into boxes of cotton waste or a special "recovery tank" filled with water), a comparison of the two is made with the comparison microscope. As described in the previous chapter, this is basically two compound microscopes joined by means of a *comparison bridge,* which consists of a system of lenses and mirrors that permits the two images to be viewed side by side, with the field of view divided equally.[34] The process of comparison works as follows:

> After the two bullets are mounted, the usual practice is for the examiner to scrutinize the entire surface of the rotating bullets at relatively low magnifications for the purpose of locating on one of the bullets the most prominent group of striations. Once such marks are located, say on the evidence bullet, that bullet is permitted to remain stationary. Then the examiner rotates the other, or test, bullet in an attempt to find a corresponding area with individual characteristics that match those on the evidence bullet. If what appears to be a match is located, the examiner rotates both bullets simultaneously to determine whether or not similar coincidences exist on other portions of the bullets. Upon finding corresponding marks on other portions, while having the bullets in the same relative positions as when the first matches were observed, the examiner proceeds with further examinations of the same nature at higher magnifications. A careful study of all the detail on both bullets ultimately permits him to conclude that both bullets were or were not fired through the same barrel.[35]

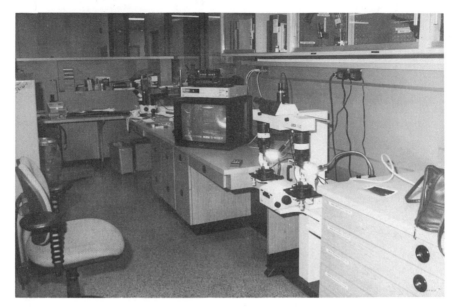

FIGURE 4.2. The workhorse of the firearms section is the comparison microscope. As illustrated here, it may be equipped with video cameras and printers, allowing for group viewing and producing photographic-quality pictures of an image. (Courtesy of Dr. John Kokanovich, Mesa Police Department, Mesa, Arizona.)

The comparison may be recorded on one or more photomicrographs. It should be cautioned, however, that

> Even if bullets were fired in succession from the same weapon, not all individual characteristics would be identical. There would be some stria-tions caused by powder residues, rust, corrosion and pitting, sand or dirt, and other surface factors or "fugitive" materials which of course are not likely to be duplicated on all bullets fired through that particu-lar barrel. Moreover, there might be other striations on the bullets which would have no relationship to the interior of the barrel through which they were fired. For instance, there might be marks on metal-cased bul-lets due to imperfections on the interior of the sizing die used in the fabrication of the bullet. Likewise, fired bullets might contain crimp or burr impressions left there by the mouth of the cartridge case or shell. Obviously, the presence or absence of such marks, whether duplicated or not, must be discounted by the firearms identification technician.[36]

If the entire process seems somewhat laborious and painstaking, it is. Therefore, firearms technicians welcome a new development in the field:

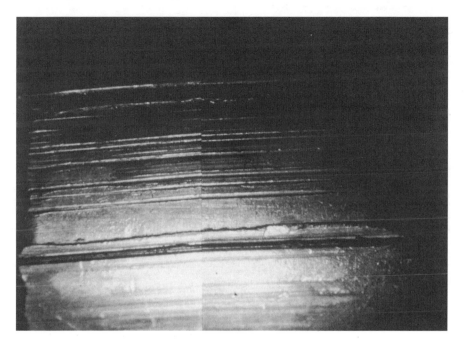

FIGURE 4.3. This split-image photomicrograph shows a match between two bullets as seen through the comparison microscope. The evidence bullet is on the left, and the test or "known" bullet is on the right. (Courtesy of Thomas Fadul, Metro-Dade Police Department, Miami, Florida.)

state-of-the-art computer systems and software that enable crime laboratories to trace bullets and shell casings through the recognition of their identifying characteristics. Programs with names such as "Bulletproof" and "Drugfire" do not actually take over the work of the firearms expert; rather, they function as sorting aids. Markings on a bullet or shell are typically recorded by microscope and video camera and then are digitally translated for computer storage. The computer can subsequently report those files with similar characteristics so that an examiner can then manually compare them and perhaps make a match. The Bulletproof database can store and retrieve data on three hundred thousand bullets and compare fifty-five thousand in under two hours—work that would require years by the manual method. It has been estimated that the program can perform the work of fifty firearms technicians.[37]

In collecting firearm evidence, care should be taken not to place accidental markings on soft lead bullets (as from the unwise use of forceps). Also, no foreign object should ever be inserted in the barrel of a gun (despite the practice of TV detectives who use a pencil to pick up a pistol

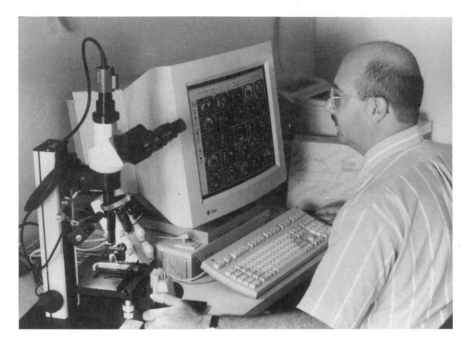

FIGURE 4.4. A computerized system known as "Drugfire" is designed to search evidence shells and bullets through a national data bank of other shells and bullets recovered around the nation. This system has had considerable success. (Courtesy of Thomas Fadul, Metro-Dade Police Dept. Miami, Florida.)

so as not to leave fingerprints). Such a practice may scratch the inside of the barrel, thus altering the striation evidence that will subsequently be produced on test bullets. Tissue, blood, fibers, or other trace evidence may be dislodged as well.[38]

Shell Comparison. The expended shell cases from fired cartridges also have markings that individualize them. However, although shell identification may constitute valuable evidence in some cases, it is much less useful than bullet identification. There are two main reasons: First, unless a shell case was discovered at the scene of a crime, the proof that it came from a particular gun can have little probative value; second, the mere fact that a shell is at a crime scene does not prove—as a bullet does—that the firearm it came from was the one used in the crime.[39]

Shell identification is based on certain markings left on the case by the firearm's mechanisms. Most of the markings are found on the base, or closed end, of the case, the end where the primer is located, and they are studied and compared in juxtaposition with the comparison micro-

FIGURE 4.5. A match illustrated by a photomicrograph taken through a comparison microscope. In this instance, the breechblock (or breech face) markings on two shells are being compared. (Courtesy of Dr. John Kokanovich, Mesa Police Department, Mesa, Arizona.)

scope. *Firing pin indentations* are produced when the firing pin is struck by the hammer and forced into the primer, leaving a crater. *Breech face markings* are caused by the burning gases inside the casing forcing the cartridge back against the weapon's breach face. Any striations on the breach face are recorded on the shell.

In semiautomatic and automatic firearms, both *extractor markings* and *ejector markings* are left by the respective mechanisms on the rim of the shell case. Also, in semiautomatic pistols, the magazine may leave marks on the side of the cartridge. And, depending on the firearm, certain additional markings may be imparted to the shell case as the result of some particular mechanism.[40]

A Florida case illustrates the potential value of shell-case evidence, as well as the value of perseverance and even inspiration on the part of detectives. The case began on New Year's Day 1991, when the body of a seminude black female was found at a wooded area on the outskirts of Apopka, Florida. Lying face down in dirt and leaves, the victim, identified as seventeen-year-old Sharonda Jackson, had been raped and shot—

an astonishing twenty-three times. For two months, the Orange County Sheriff's Department had only false leads in the case.[41]

Then there were two attempted abductions at gunpoint, but in each case the women were able to break free and escape. The police soon identified a thirty-nine-year-old African American named Robert Jerome Perkins as the assailant, and he was arrested. On March 22, detective Stuart DeRidder revisited the murder site. Sifting through the sandy loam, he and another detective eventually uncovered three 9mm bullets and seven spent 9mm brass shells stamped "W.C.C. 89" (meaning Winchester Cartridge Corporation, 1989). DeRidder recalled that a woman who had been raped two weeks before Sharonda Jackson's murder had been shot at during her abduction. She was questioned again by detectives, and she identified Robert Perkins from a photo lineup of five men of similar appearance. The detectives next took a metal detector to the site of that woman's abduction, where they recovered a 9mm casing stamped "W.C.C. 89."

Lacking the suspect's handgun, the detectives next went to his residence where, with permission of his wife, they scanned the yard with the metal detector. Soon they found a spent shell with the "W.C.C. 89" stamping. Laboratory examination by Susan Komar and Gary Rathmann of the Florida Department of Law Enforcement showed that all of the shell casings had been fired by the same gun. Finally, DNA analysis showed that semen taken from Sharonda Jackson's body matched a sample of blood taken by court order from Robert Perkins. A first trial ended in a hung jury, but on retrial Perkins was convicted of the first-degree murder, kidnapping, and rape of Sharonda Jackson. He was sentenced to death.[42]

Serial Number Restoration. Sometimes firearms are recovered with their serial numbers ground off to conceal the owner's identity. If the grinding has not been excessively deep, the numbers can usually be restored by the examiner. Restoration depends on the fact that the underlying metal crystals are placed under a strain when the numbers are stamped into the firearm's frame. When an etching agent is applied—for example, on steel surfaces, a solution of hydrochloric acid (120ml), copper chloride (90gr), and water (100ml) is used—the strained area will dissolve faster than the surrounding metal, permitting the pattern of the stamped numbers to appear, whereupon they must be promptly photographed. Other manufactured items, including engine blocks of automobiles, may also have serial numbers that require restoration.[43] Depending on the metal, other techniques, such as heating or electrochemical etching, may be effective.[44]

FIREARM RESIDUES

The explosion that occurs when a firearm is discharged produces two types of trace evidence: gunshot residues and powder pattern deposits.
Gunshot Residue Analysis. Particles of gunshot residue (GSR) are deposited on the hand or hands of a person firing a gun. These may adhere for up to six hours or so, but they may also be removed by washing or wiping the hands. Obviously they are present in the greatest concentration immediately after the shooting, usually in the web area between the thumb and forefinger.[45]

There are three methods of GSR testing and, depending on which method is used, two methods of collection. If the testing is to be by either neutron-activation analysis (NAA) or atomic absorption spectrophotometry (AA), the evidence is collected by using cotton-tipped plastic swabs (wood contains substances that interfere with the test). A 5 percent solution of nitric acid is sprayed on the hands, then the areas are swabbed twice. Each set of swabs is placed into a plastic tube (glass may also interfere with the test results). Tubes are labeled appropriately: "left back," "right back," "left palm," "right palm," and "control"; a single swab is used for the cartridge. Special evidence collection kits are generally used for the purpose.[46]

For testing by scanning electron microscopy/energy dispersive x-ray analysis (SEM/EDX), small aluminum discs to which double-sided cellophane tape is affixed are used for collection. These also come in kit form. The adhesive is pressed on the top of the webbed portion of each hand, with one disc used for each hand.[47] According to *Techniques of Crime Scene Investigation*:

> A word on the interpretation of test results is called for. A negative result may mean that a subject either did *not* fire a weapon or took some action that resulted in removal of any particles that had been present; a positive test result means that the subject fired a weapon sometime during the past 6 hours (approximately), or handled a weapon during that time period. The location of the particles, of course, may be important. If the palms, but not the backs, of a subject's hands contain GSR, this might indicate that the person held but did not fire a weapon, or that the weapon was relatively "clean," i.e., gave off a small amount of GSR particles when fired.[48]

The old "paraffin test" or "dermal nitrate test" is no longer recommended by authorities. The test lacks specificity and common materials such as urine, tobacco, cosmetics, and fertilizer give false positive results.[49]

FIGURE 4.6. A high-speed photograph shows the discharge residue from a revolver shortly after the bullet has left the barrel. The residue may comprise numerous chemicals, most significantly lead, barium, antimony, nitrites, and carbonaceous smoke. (Courtesy of Thomas Fadul, Metro-Dade Police Department, Miami, Florida.)

Powder Pattern Deposits. When a firearm is discharged at relatively close range, gunpowder and other materials may be spewed onto the target. These deposits may include burned and unburned particles of gunpowder, nitrates, lead and barium, and other chemicals. Factors such as the type of gun and length of barrel, as well as the type of ammunition, will affect the powder pattern—both in size and in density. The appearance of the pattern may help indicate the distance between firearm and target. If the weapon is fired a half inch or closer to the target, there usually is no powder pattern. Tests with the same firearm and ammunition fired into heavy paper or cardboard at varying distances should give a range of patterns that can then be compared to the evidence to estimate distance.[50]

If the weapon has not been recovered, "the best that the examiner can do," advises one expert, "is to state whether or not a shot could have been fired within [the] same distance interval from the target."[51]

In the case of garments, the surface should be examined microscopically for traces of powder deposits. Sometimes, because of the color of the clothing or the presence of blood, the particles of deposit will not be apparent. In such cases an infrared photograph may overcome the problem, or chemical tests may be used to detect the chemical residues.[52]

CASE STUDY: THE SACCO AND VANZETTI TRIAL

The event that sparked one of the most controversial cases in American history took place at the beginning of the "Roaring Twenties" about forty miles from Salem, Massachusetts (scene of the tragic "witch" trials of the seventeenth century). On Friday, April 15, 1920, the Slater and Morrill Shoe Factory in South Braintree, Massachusetts, was awaiting delivery of its weekly payroll—a sum of sixteen thousand dollars. Paymaster Frederick A. Parmenter was carrying a black payroll money bag. A similar bag was carried by an armed guard named Alexander Berardelli. Neither man seemed to pay any attention to two loiterers leaning against a fence near the Pearl Street factory.

Suddenly, as the payroll carriers passed the two men, one of the loiterers sprang forward and, pulling a pistol from his pocket, shot Berardelli at close range. As the guard fell, Parmenter dropped his money bag and fled across the street. The gunman pursued him, bringing him down with two shots. While Parmenter was being chased and shot, his accomplice stooped over the wounded guard and fired three more bullets into him. They gathered up the two money bags, a large black Buick came careening around a corner, and the robbers jumped in. They had completed their work in a few brief minutes. They left behind two fatally injured men. The guard died at once, and the paymaster expired later.

Evidence found immediately at the scene consisted of six ejected .32-caliber cartridge cases and some eyewitnesses. According to them, there had been a third gunman at the scene, although he had not participated in the robbery. He had leapt with the other men into the getaway car, which had contained two additional accomplices.

Of the five, only the two assailants were described in detail. They were "foreign-looking," one clean-shaven, the other with a heavy mustache. The former was the one who first shot Berardelli and then pursued and fatally shot Parmenter. Soon firearms experts would determine that this man had used a Colt .32 automatic (which had ejected its cartridges as they were spent).[53]

The first lead in the case came a few days later with the discovery of the getaway car, which had been abandoned in a wooded area. A man named Boda, who had been seen in the car before it was abandoned, was traced to another car that was undergoing repair in a West Bridgewater garage. When Boda arrived to pick up the repaired car, he became frightened and fled on a motorcycle before police could close in. He later escaped to his native Italy. However, two other men who had been with him were apprehended later on a streetcar.[54] The suspects matched the descriptions of the killers. They were Italian immigrants, a clean-shaven factory worker, Nicola Sacco, age twenty-nine, and a heavily mustached Bartolomeo Vanzetti, thirty-two years of age, a fish peddler. Both were away from their homes in Plymouth; both were armed. Found in Sacco's pocket was a .32 Colt automatic containing a fully loaded magazine; he also had extra cartridges, variously manufactured by Peters, Remington, and Winchester. Vanzetti carried a loaded .38-caliber revolver. In addition he had some shotgun shells similar to one left at the scene of an earlier—abortive—payroll robbery in Bridgewater.

That the men were at least petty criminals was attested to by the contents of their pockets; they were illegally in possession of the firearms, and one of the handguns was of the same make and model as that used in the shoe factory payroll robbery. (Later, the men claimed the weapons were needed for their self-protection because they lived, they said, in such "bad times.") But there was more: The suspects were also in possession of what was called "anarchist literature." One tract taken from Sacco began, "Fellow workers," and decried "all the capitalists." Before the duo could be tried for the double murder, Sacco was convicted of the earlier Bridgewater robbery attempt, although Vanzetti had an alibi for that time.[55]

Ominously, the same judge who presided at Sacco's trial, Webster Thayer, was also the judge at the trial of the two immigrants, which opened on May 31, 1921, at Dedham, Massachusetts. Emotions ran high, and the defendants received much sympathy. According to one account of the case:

> Sacco and Vanzetti were poor Italian immigrants who spoke little English. But both men were anarchists, members of an international movement that believed that true social justice for the poor and oppressed could be achieved only by the destruction of established governments. In America, anarchists had purportedly been responsible for numerous bombings, among them attempts on the life of John D. Rockefeller and the attorney general of the United States. The government had responded with mass arrests and deportations. So what began as a local robbery and murder case grew into an international political event, and Sacco and Vanzetti were seen as martyrs in the struggle of the downtrodden against the rich and powerful. Throughout the world many

people believed that their arrest was simply a crude effort by the ruling class to crush political dissent.[56]

For their part, the agents of the "ruling class" did little to allay fears that they might be prejudiced against the defendants. As another source observes: "Those wishing to retain a due confidence in the Olympian objectivity of the law have found it an edifying experience to overhear the comments of the officials on the case. Chief prosecuting attorney Frederick Katzmann regarded them as 'those damned God-hating radicals,' while Judge Webster Thayer's locker-room style boast after the trial was 'Did you see what I did with those anarchist bastards the other day?'"[57] Approximately sixty witnesses testified for the prosecution and nearly one hundred for the defense. Much of the testimony concerned the question of identity and was conflicting.

Then came the firearms evidence. In all, four pistol bullets were recovered from the body of the guard, Berardelli; one from the body of the paymaster, Parmenter; and another from the latter's clothing—all .32-caliber. The prosecution's lead witness was Massachusetts State Police Captain William Proctor, who was "doubling" as a firearms expert. He described how he and a colleague had test-fired fourteen bullets from Sacco's Colt automatic into a special bullet-recovery box filled with oiled sawdust. When they compared these bullets with the six recovered from the victims, the rifling marks on the test bullets matched those on only one of the questioned bullets, that determined to have caused Berardelli's death; the other five presumably had been fired from another pistol. When asked by the assistant district attorney if the fatal bullet had been fired from Sacco's pistol, Proctor replied, "My opinion is that it is consistent with being fired by that pistol."[58] Regarding the mortal bullet, Proctor's colleague, Charles Van Amburgh, stated to the court, "I am inclined to believe that it was fired from this [Sacco's] automatic pistol." Amburgh added that rust at the bottom of the pistol barrel could have produced the traces of rust found on the death bullet.

This evidence was clearly equivocal and was challenged by the experts for the defense. The first of these was James Burns of the U.S. Cartridge Company, who had some thirty years' experience as a ballistics engineer. He thought the fatal bullet had been fired by a Bayard, a pistol of foreign manufacture. Asked specifically whether it came from Sacco's automatic, he replied: "In my opinion, no. It doesn't compare at all."

Burns was supported by the second defense firearms witness, J. Henry Fitzgerald, head of the testing room at the Colt Patent Firearms Company.

He testified, "I can see no pitting or marks on the bullet that would correspond to a bullet coming from this gun." Obviously the defense testimony was also equivocal. While it is clear that all the firearms testimony came from men who had a certain knowledge of arms and ammunition, it is equally clear that none of them had training as firearms examiners, with the possible exception of Captain Proctor, who knew something of the class characteristics of rifling.[59] As one writer summed up the testimony concerning whether the fatal bullet was fired by Sacco's pistol: "Two self-taught 'experts' testifying for the prosecution said the marks proved it had been; two self-taught 'experts' testifying for the defense said it had not."[60]

Interestingly, during jury deliberations in the case, the jurors asked to be supplied with a magnifying glass so that they could examine the bullet for themselves. Apparently they concluded that the fatal bullet had indeed been fired from Sacco's pistol. The jury found the defendants guilty of murder in the first degree, and they were sentenced to execution.

The death sentence, however, rather than being the conclusion of the case, was in reality only the beginning of a seven-year period of controversy. The defendants' plight became a *cause célèbre* worldwide. Large demonstrations on their behalf were held in Moscow and Paris as well as in major cities in South America. Requests for their release came from the British Labour Party, the German Reichstag, and the French Chamber of Deputies. Around the world, American embassies were bombed and U.S. officials threatened. The artistic community responded emotionally, producing books, paintings, and plays on behalf of the condemned men.[61]

Indeed, the case still attracts conspiracy-minded theorists who perpetuate the myth that Sacco and Vanzetti were innocent victims of perverted justice. Michael Kurland's *A Gallery of Rogues: Portraits in True Crime* heralds the entry on the case with the heading: "Two Gentle Anarchists Convicted and Executed for a Crime They Probably *Did Not* Commit." Kurland points to Captain Proctor, who "later admitted in an affidavit that his statement that one of the murder bullets had markings 'consistent with being fired by that pistol' merely meant that the bullet could have come from Sacco's gun, or many others, and that he personally did not believe that it had been fired by Sacco's gun. He further admitted that he had been heavily coached by the prosecution."[62] Kurland goes on to point out that many years later mobster Vincent Teresa claimed he had been told by Butsey Morelli (of the notorious Morelli brothers, founders of the New England branch of the Mafia) that he and his brothers had robbed the shoe factory payroll. "These two suckers took it on the chin for us," Kurland quotes Butsey as saying. "That shows you how much justice there really is." (Elsewhere in the same book, Kurland again quotes Butsey's alleged statement, only this time "suckers" is replaced by "greaseballs"—a variant in the quotation that hints at folklore in the making.)[63] Of course, if matters were as Kurland represents them, one might accept his conclusion, but there are crucial developments in the case that he has managed to overlook.

During the years while the case was making international headlines, important advances were being made in the field of forensic firearms comparison. In 1920, Charles E. Waite, investigator in the office of the New York state prosecutor, began to collect data on firearms. Within five years, he had catalogued details of all American and European firearms, including the caliber, direction of rifling twist, and number of grooves. He came to realize that the markings in barrels should be individual because of the wearing of the machine tools used in producing them. Unfortunately, he had no effective means of examining the interiors of gun barrels.

Waite explained his problem to John H. Fischer, a physicist who had developed a medical instrument called the cystoscope (a tubular probe that permits viewing of the bladder and kidneys without surgery). In response, Fischer invented a similar device called the *helixometer*—a long, hollow tube with a light and magnifier at the tail end. "With a few modifications," reports one authority, "the helixometer is still a fixture of every firearms laboratory."

Waite also sparked the interest of chemist and microphotographer Philip O. Gravelle, who photographed by the thousands bullets that had been test-fired into recovery containers filled with cotton wool. He observed that each bullet contained markings—striations—that were unique to each weapon—its "fingerprint," so to speak. Gravelle then proceeded to invent an optical device that could link two compound microscopes so that two bullets could be examined simultaneously—the comparison microscope. The side-by-side comparisons it permitted helped remove human error from the examining process.[64]

In 1923, together with Fischer and Gravelle, Waite founded the Bureau of Forensic Ballistics in New York City. Two years later, Waite added another member to the team, a physician named Calvin Goddard (whose pioneering work in the field has been mentioned earlier). Goddard was a former major in the Army Medical Corps who had risen to become deputy director of Johns Hopkins Hospital. Goddard brought to the bureau not only his medical knowledge but also a great fascination with firearms. Upon Waite's death from a heart attack in 1926, Goddard became the team's leader. The following year he became involved in the Sacco and Vanzetti case.[65]

Goddard entered the case in 1927 after Massachusetts governor A.T. Fuller had acquiesced to international pressure and appointed a three-man commission to review the facts and advise him whether he should grant Sacco and Vanzetti clemency. Headed by Lawrence Lowell, president of Harvard University, the Lowell Commission carefully reviewed the case and—except for commenting on Judge Thayer's "grave breach of official decorum"—found that the defendants had received a fair trial and were convicted on the evidence.[66] Goddard's offer of free assistance came too late to be included in the report, but he received permission to set up his comparison microscope and helixometer in the Dedham Courthouse. Present were experts and counsel for both sides, and four newspaper reporters.[67]

Goddard first examined the shells from the scene of the crime, using techniques he had developed that soon became standard practice in crime laboratories. Goddard placed a shell found near Berardelli's body under one objective of his comparison microscope, and under the other positioned a shell test-fired from Sacco's gun. They did not match, proving that Sacco's pistol did not fire that round. The same was true of a second bullet from the scene of the murders. However, when Goddard placed a third shell casing from the crime under the microscope, he saw that it definitely came from Sacco's pistol. Not only had the firing pins left impressions that were exactly the same diameter, they had also left tiny V-shaped scratch marks on both primers. In addition, there were dozens of alternating furrows and ridges across the base of each shell, stamped there by the breech-block. Goddard positioned the shells so that the features of their opposite halves were matched at the center line of the microscope's viewing area. Defense attorneys objected that these scratches and ridges were not in identical positions on the two shells, but were nearly centered on one while being much more off-center on the other. Goddard explained that all firearms have some "play" in them and that shells are not always forced back against the breech block in precisely the same position. Neither does the firing pin always strike the primer at dead center. The important matter, Goddard demonstrated, was that the individual markings were identical in relationship to the whole.

Next, Goddard tested the four bullets recovered from Berardelli's body during autopsy. He compared each in turn with test bullets he had fired from Sacco's gun into a box filled with sawdust. The first bullet, recorded in the evidence as fatal bullet number three, was somewhat corroded and otherwise damaged by its contact with the victim's body. Nevertheless, the comparison microscope proved conclusively that it had been fired from the gun taken from Nicola Sacco. Not only were the rifling grooves the same width, depth, and pitch (class characteristics, consistent with a .32 Colt automatic), but there were also identical longitudinal scratches or striations (individual characteristics). In particular, the microscope revealed one prominent scratch or gouge on this bullet that exactly matched one on the test bullet. Lining up the striations from each bullet, Goddard invited one of the defense experts, James Burns, to take a look. After a long silence he remarked, "Well, what do you know about that?" Then he resigned from the defense team.[68]

Goddard's report was to be used at any new trial for the anarchists/murderers, but they had already been convicted and their convictions upheld by the Lowell Commission. Henry Morton Robinson, in his 1935 book, *Science Catches the Criminal,* stated that "Goddard's uncontroverted findings are believed . . . to have bolstered the Lowell report with the strongest kind of scientific documentation. From this time forward forensic ballistics became a new type of judicial proof which ruled out mere opinion and did much to halt the charlatanism that had prevailed in the field of firearms identification."[69]

On August 23, 1927, while millions of sympathizers protested worldwide, Nicola Sacco and Bartolomeo Vanzetti were executed along with another death-row inmate. They were electrocuted in the Massachusetts electric chair soon after midnight on that Tuesday morning. "All Reject Religious Consolation to the Last," reported the *Boston Daily Globe* in one of its sub-headlines. Vanzetti forgave those who, he still maintained, were executing an innocent man. Sacco, who had preceded him in death, was defiant to the end, condemning the social order that punished him.[70]

Robinson added these thoughts to his book:

> To adherents of the Sacco-Vanzetti cause, it may be some consolation to realize that the testimony concerning the identity of the fatal bullet proved only that Sacco's weapon was used in the commission of the crime. Goddard did not state that Sacco pulled the trigger. There remains the remote possibility, as is true in any case, that someone other than Sacco did the killing. In other words, an expert who testifies that the fatal bullet comes from a particular weapon is obviously not in a position to say who actually did the shooting. It remains for evidence of a different nature to place the smoking weapon in the hands of the

guilty individual. Nevertheless, it looks more than slightly suspicious whenever a weapon used in a commission of a crime is found in the pocket of the accused.[71]

Robinson wrote that he once asked Colonel Goddard, "If Sacco and Vanzetti had retained you as their expert how could you have reconciled your findings in this case with the best interests of your clients?" Robinson says his reply was characteristic. "'Such reconcilings and jugglings of evidence do not interest me,' he said bluntly. 'One of my earliest determinations was never to take an unjust case either for prosecution or defense. As far as I know I never have. I make my examinations and clearly state what I find. For this I usually expect a fee of one hundred dollars a day, plus expenses. If my client decides not to use my report, that is his privilege.'"[72]

Goddard went on to further forensic success in the infamous case of the St. Valentine's Day Massacre in 1929. On that date, two men disguised as police officers strode into a Chicago garage with drawn guns, placed members of Bugs Moran's bootlegging gang against a wall, and stood by while two other men blasted them to death with .45-caliber Thompson submachine guns. Ten months later, the real police, investigating a man who had shot a fellow officer, discovered a cache of weapons in his closet. Among them were two .45-caliber Thompsons. Goddard established that they were the lethal weapons in the February 14 massacre, and his testimony was crucial in obtaining a conviction that sent one of the killers to prison. (Two more suspects were killed in "revenge" slayings.) Goddard's success was rewarded by two wealthy businessmen who had served on the coroner's jury and were so impressed that they financed Goddard's own Scientific Crime Detection Laboratory at Northwestern University. He later helped the FBI set up its firearms section when its Criminological Laboratory, as it was then known, was opened in 1932. Its very first piece of laboratory equipment was a comparison microscope.[73]

Almost three decades went by before there came yet another development in the Sacco and Vanzetti case. In 1961, a panel of forensic firearms experts reexamined Sacco's pistol and one of the murder bullets and concluded that that bullet had definitely been fired from that gun. But the case had passed into legend by then. As one writer put it, "This plain scientific fact should end all controversy, but of course it will not. It is in the nature of things that Sacco and Vanzetti will always be remembered as folk heroes in the fight for freedom, rather than as brutal killers who murdered for gain."[74]

NOTES

1. Brian Marriner, *On Death's Bloody Trail* (New York: St. Martin's Press, 1991), 129-30.

2. Barry A.J. Fisher, *Techniques of Crime Scene Investigation*, 5th ed. (New York; Elsevier, 1992), 272.

3. "Firearms Identification," forensic instruction manual for course in Scientific Crime Detection (Chicago: Institute of Applied Science, 1962), lesson 2, 4.

4. Lucien C. Haag, *Forensic Firearms Evidence: Elements of Shooting Incident Investigation* (Pinole, Calif.: ANITE Productions, 1991), 4-6.

5. W. Emerson Reck, *A. Lincoln: His Last 24 Hours* (Columbia: Univ. of South Carolina Press, 1987), 102, illus., 97.

6. Rowe, Walter F., "Firearms Identification," in *Forensic Science Handbook*, vol. 2, ed. Richard Saferstein (New York: Prentice-Hall, 1982), 395.

7. Fisher, *Techniques*, 272, illus., 273.

8. C.B. Colby, *F.B.I.: The 'G-Men's' Weapons and Tactics for Combating Crime* (New York: Coward-McCann, 1954), 16.

9. Haag, *Forensic*, 9-13; Rowe, "Firearms," 395.

10. Haag, *Forensic*, 9-13; Rowe, "Firearms," 395.

11. Fisher, *Techniques*, 272; Rowe, "Firearms," 395-96.

12. Fisher, *Techniques*, 273; Rowe, "Firearms," 395-96; "Firearms Identification," lesson 3, 13.

13. Fisher, *Techniques*, 273-74; Rowe, "Firearms," 397-98.

14. Fisher, *Techniques*, 274-75; Rowe, "Firearms," 404-6.

15. Fisher, *Techniques*, 275. Rowe, "Firearms," 401-3.

16. Haag, *Forensic*, 39.

17. "Firearms Identification," lesson 4, 14-16.

18. Ibid., lesson 1, 4.

19. Ibid., 10; Rowe, "Firearms," 394.

20. "Firearms Identification," lesson 1, 12.

21. Ibid., 12-17; Rowe, "Firearms," 394.

22. "Firearms Identification," lesson 1, 17.

23. Ibid., 17-18.

24. John F. Fischer and Joe Nickell, "'Keyhole' Skull Wounds: The Problem of Bullet-Caliber Determination," *Identification News* (Dec. 1986): 8-10.

25. Joe Nickell with John F. Fischer, *Mysterious Realms: Probing Paranormal, Historical, and Forensic Enigmas* (Buffalo, N.Y.: Prometheus Books, 1992), 107-29.

26. Ibid., 122.

27. Philip Paul, *Murder under the Microscope: The Story of Scotland Yard's Forensic Science Laboratory* (London: Futura, 1990), 57. Another source reverses the ridge/groove features: see Charles R. Swanson Jr., Neil C. Chamelin, and Leonard Territo, *Criminal Investigation*, 4th ed. (New York: McGraw-Hill, 1988), 18-19.

28. Swanson, Chamelin, and Territo, *Criminal Investigation*, 19; Marriner, *On Death's Bloody Trail*, 134-35; Paul, *Murder under the Microscope*, 58.

29. Frank Smyth, "Ballistics," in *Science against Crime,* ed. Yvonne Deutch (New York: Exeter Books, 1982), 138; Swanson, Chamelin, and Territo, *Criminal Investigation,* 19.

30. Swanson, Chamelin, and Territo, *Criminal Investigation,* 19.

31. Marriner, *On Death's Bloody Trail,* 135.

32. "Firearms Identification," lesson 6, 10; Fisher, *Techniques,* 275; Swanson, Chamelin, and Territo, *Criminal Investigation,* 101.

33. Rowe, "Firearms," 421-23.

34. Ibid., 424; Fred E. Inbau, Andre A. Moenssens, and Louis R. Vitullo, *Scientific Police Investigation* (New York: Chilton, 1972), 75-76.

35. Inbau, Moenssens, and Vitullo, *Scientific Police Investigation,* 76-77.

36. Ibid., 77-78.

37. Katy Benson, "Match Makers," *Police* (June 1995): 32-35; Tod W. Burke, "Drugfire and Bulletproof," *Law and Order* (July 1995): 99-101.

38. Swanson, Chamelin, and Territo, *Criminal Investigation,* 105.

39. Inbau, Moenssens, and Vitullo, *Scientific Police Investigation,* 79.

40. Swanson, Chamelin, and Territo, *Criminal Investigation,* 102; "Firearms Identification," lesson 7, 1-6.

41. Sam Roen, "Florida's Happy-New Year Rape Slayer," *Startling Detective* (July 1995): 37-38.

42. Ibid., 38-41.

43. Richard Saferstein, *Criminalistics: An Introduction to Forensic Science,* 5th ed. (Englewood Cliffs, N.J.: Prentice Hall, 1995), 456; Fisher, *Techniques,* 291.

44. Fisher, *Techniques,* 291.

45. Ibid., 277-78.

46. Ibid.

47. Ibid., 279.

48. Ibid.

49. Saferstein, *Criminalistics,* 453.

50. Fisher, *Techniques,* 302-3.

51. Saferstein, *Criminalistics,* 448.

52. Ibid., 451.

53. Smyth, "Ballistics," 144-45. Except as otherwise noted, information on the Sacco and Vanzetti case was taken from this source.

54. Marriner, *On Death's Bloody Trail,* 141 (this source reports incorrectly that Sacco was carrying a .32 revolver); David Fisher, *Hard Evidence* (New York: Dell, 1995), 289.

55. Smyth, "Ballistics," 145; Martin Fido, *The Chronicle of Crime* (New York: Carroll E. Graf, 1993), 152.

56. Fisher, *Hard Evidence,* 289.

57. Smyth, "Ballistics," 146.

58. Ibid.; Fisher, *Hard Evidence,* 290.

59. Smyth, "Ballistics," 146.

60. Fisher, *Hard Evidence,* 290.

61. Ibid.

62. Michael Kurland, *A Gallery of Rogues: Portraits in True Crime* (New York: Prentice Hall, 1994), 336.

63. Ibid., 337. See also 267, 375-76.

64. Paul, *Murder under the Microscope*, 58-59; Smyth, "Ballistics," 138-42.

65. Paul, *Murder under the Microscope*, 59; Smyth, "Ballistics," 142.

66. Kurland, *A Gallery of Rogues*, 337; Henry Morton Robinson, *Science Catches the Criminal* (New York: Blue Ribbon, 1935), 85.

67. Robinson, *Science Catches the Criminal*, 83.

68. Ibid., 83-85.

69. Ibid., 85-86.

70. *Boston Daily Globe*, Aug. 23, 1927.

71. Robinson, *Science Catches the Criminal*, 86.

72. Ibid.

73. Fisher, *Hard Evidence*, 292-93; Paul, *Murder under the Microscope*, 59-60.

74. Marriner, *On Death's Bloody Trail*, 143.

RECOMMENDED READING

Guinn, V.P. "JFK Assassination: Bullet Analyses." *Analytical Chemistry* 51 (1979). A presentation of results of neutron activation analyses that compared the bullet taken from Gov. John Connally's stretcher with bullet fragments recovered from the car, Connally's wrist, and President John F. Kennedy's brain. There was evidence of only two bullets, consistent with Warren Commission findings.

Haag, Lucien C. *Forensic Firearms Evidence: Elements of Shooting Incident Investigation.* Pinole, Calif.: ANITE Productions, 1991. A reference handbook (with accompanying two-volume videotape set) intended to provide the basic elements of the reconstructive aspects of firearms examination.

Lichtenberg, L. "Methods for the Determination of Shooting Distance," *Forensic Science Review*, vol. 2, no. 1 (June 1990). A description of advantages and disadvantages of the various scientific methods of determining shooting distance.

Rowe, Walter F. "Firearms Identification." In *Forensic Science Handbook*, vol. 2, ed. Richard Saferstein. Englewood Cliffs, N.J.: Prentice Hall, 1988. Forensic text on fundamentals of firearm examination, including arms and ammunition, history, examination of bullets and fired cartridges, and estimation of range of fire.

Smyth, Frank. "Ballistics." In *Science against Crime*, ed. Yvonne Deutch. New York: Exeter Books, 1982. Readable, popular text on firearms examination including history, ballistics, rifling marks. Covers three case studies including the Sacco and Vanzetti case.

5 FINGERPRINTING

Great interest was shown them in ancient times, then for centuries they were neglected, but fingerprints have played important roles in modern crime detection. For example, after a Mannlicher-Carcano rifle was recovered following the assassination of President John F. Kennedy, the Dallas Police crime lab found underneath the stock a latent palmprint that belonged to Lee Harvey Oswald.[1] In 1968, latent fingerprints on a rifle led to the arrest of James Earl Ray for the assassination of Dr. Martin Luther King Jr.[2] And that same year fingerprints established the identity of the young alien whom police had in custody following the murder of Senator Robert F. Kennedy but who refused to say who he was: Sirhan Bishara Sirhan.[3] As one forensic text puts it, "Today the fingerprint is the pillar of modern criminal identification."[4] Because the topic of fingerprinting is rather broad, we will not discuss in great detail such matters as differentiation of pattern variants and filing of prints but will focus instead on the *history* of the science, as well as fingerprint *classification* and *identification*, and *processing and recovery.*

HISTORY

The earliest recognition of the uniqueness of fingerprints and their suitability for personal identification apparently came from the ancient Chinese, who employed a thumbprint in lieu of a signature on legal conveyances and even criminal confessions. Since literacy was uncommon, this proved a practical measure. The first scientific recognition of fingerprints in the West came from certain late-seventeenth-century writings. In 1684, Dr. Nehemiah Grew issued a report before London's Royal Society wherein he described the ridges and pores of the hands and feet. The following year, one G. Bildoo published a treatise on anatomy that described the sweat pores and textured ridges, and the next year a professor of anatomy at the University of Bologne, Marcello Malpighi, referred to the "varying ridges and patterns" of human fingertips.[5] He did not further comment on their function or utility, except to state that the ridges were "drawn out into loops and whorls."[6]

The first person to devise a system of classifying fingerprints was a Prussian professor, Johannes E. Purkinje (1787-1869) of Breslau University. In 1823, he published a thesis, in Latin, in which he described nine fingerprint pattern types, gave each a name, and set down rules for their individual classification—many of which are still followed today. Neither he nor apparently any of the others mentioned here, however, ever suggested that fingerprints might be used for identifying people.[7] That recognition came from two other men—one in India, the other in Japan—neither of whom was aware of the other's work.

In 1858, William Herschel (1833-1917), a grandson of noted British astronomer Sir William Herschel, was the chief administrator of the Hooghly District of Bengal, India. He was confronted with frequent impersonations among natives attempting to obtain pensions and other allowances, so he began recording the handprints of the Bengalese on contracts to prevent impersonation or subsequent repudiation of signatures. In 1877, Herschel sought to implement his fingerprint system in the Bengal jails but could not obtain permission.

In the meantime, Dr. Henry Faulds (1843-1930), a Scottish physician and surgeon who was practicing in Taukiji Hospital in Tokyo, was also conducting research into fingerprints. Like Herschel, he had discovered that oil and sweat from the pores resulted in latent (invisible) prints that could be developed with powders. Faulds utilized this technique to exonerate a man accused of burglary by showing that the fingerprints that had been found beside the window belonged to another man, who was later caught. Thinking he was the first pioneer in fingerprinting, Faulds published his research in the October 28, 1880, issue of *Nature* magazine. This provoked a letter to the editor from Herschel, who asserted his claim. Faulds appealed to Charles Darwin, the great naturalist, who referred him to his cousin, Sir Francis Galton (1822-1911), a researcher in heredity.

Galton proved less interested in settling the dispute than in joining in the interesting new field. He published a textbook, *Finger Prints*, in 1892, giving credit to Herschel as "the first to devise a system for the use of fingerprints for identification" and, most important, setting forth a practical method of classification—not just by patterns but by the entire set of fingerprints of a given individual so that they could be filed.[8] Galton also determined that fingerprints were unchanging over time, but he was not the first to do so. Such proof was first offered by a German researcher, Herman Welcker. In 1856 he had made a print of his right palm, and he repeated the process in 1897 and published both impressions, showing that, despite a forty-one-year period between the two prints, the ridge patterns had not changed.[9]

In the meantime, Herschel's successor in India, Edward Henry (1850-1931), became interested in fingerprinting, and he began work on his own classification system. Where possible he simplified, where necessary he added complexity, and overall he shaped the whole into a fully practical and workable system. Completed in 1899, his system was published in book form the following year. In 1901, the Henry system of fingerprint classification and filing was implemented in England and Wales. With later modifications by the FBI, the Henry system continues in use to the present day. (In Spanish-speaking countries of Central and South America, a different system is used, one that was devised by Juan Vucetich [1858-1925], an Argentinian police official, and was also based on Galton's work.)[10]

The year after Galton's book appeared, the British Government appointed a group known as the Asquith Committee to investigate the subject of fingerprinting. The committee was impressed with what it saw in Galton's laboratory, and its report of February 12, 1894, resulted in the official adoption in the United Kingdom of fingerprinting as a supplementary identification system.[11]

The primary method of identification at that time was *bertillonage*, a system of anthropometry (the science of measuring the human body), named for its creator, Alphonse Bertillon (1853-1914). Bertillon's method involved recording eleven measurements: the length of the left arm (from elbow to tip of middle finger), length and breadth and diameter of skull, width of outstretched arms, height sitting and standing, length of left middle and little fingers, length of the left foot, and length of right ear, in addition to other physical characteristics (eye color, scars, and so on), plus profile and full-face photographs. Bertillon calculated that the variations in just the eleven measurements in his system gave odds better than 286 million to one against two individuals having the same measurements. Implemented in France in 1883, Bertillon's system in only one year identified some three hundred criminal recidivists that the police detectives had missed. Although bertillonage put identification on a scientific basis, it was slow and cumbersome, and different identification officers could obtain slightly different measurements.[12]

By the turn of the century, Scotland Yard had begun to rely on fingerprints. Reluctantly, Bertillon added space on his record card for such prints—of the right hand only. But in 1903 in the United States came the case that marked the beginning of the end for bertillonage. On May 1 of that year, a young African American man named Will West was admitted to Leavenworth Penitentiary at Leavenworth, Kansas, and his Bertillon measurements were taken. The records clerk thought he looked familiar, although West denied having been there before. When the files were checked, a card was discovered with similar measurements and bearing photographs of what looked like the same man. The name on the card was "William West." After officials got over their initial confusion, they brought together in the same room Will West and William West, who looked as alike as twins, although they were said to be unrelated. This was just the case fingerprint advocates were looking for. They quickly demonstrated that they could infallibly distinguish one man from the other.[13] As a result, Leavenworth went over to the new system, reportedly "the next day" (and the following year a national fingerprint bureau, to include all federal prison inmates, was established there).[14]

That is essentially the story of "The Two Will Wests," which helped sound the death knell of an already dying bertillonage. In fact, however, far from representing "one of the strongest coincidences in all history," as one source asserted,[15] the events were somewhat less dramatic than they have been represented. As one of us (J.N.) discovered in 1980 and published in *The Journal of Police Science and Administration*, the Wests were actually identical twins. Proof of that fact included genetic evidence from similar fingerprint patterns on each hand and similar ear configurations (even fraternal twins would not have had such similar ear patterns), plus similar biographical details, the testimony of a fellow prisoner who knew the men to be "twin brothers," and the fact that, while in Leavenworth, both Bill and Will West wrote to the same brother, the same five sisters, and the same Uncle George (as shown by the correspondence log).[16]

In any event, New York's Sing Sing Prison—which went over to fingerprinting only a month after the events at Leavenworth—would probably have done so anyway. The U.S. Army followed suit in 1905, followed by the Navy in 1907, and the Marine Corps in 1908.[17] By 1915 fingerprint technicians were so numerous that they created the International

Association for Identification, and in 1919 the first journal of the profession, *Finger Print and Identification Magazine*, began publication. In 1924 the U.S. Congress established at the then-Bureau of Investigation (now FBI) the Identification Division, which was to serve as the nation's repository of all fingerprint records.

During the gangster era of the twenties and thirties, fingerprinting captured the imagination of the American public when it was seen as the premier scientific weapon for dealing with public enemies.[18] Indeed, Public Enemy Number One, John Dillinger, sought to have his fingerprints altered at the same time he underwent minor plastic surgery to change his face. The result was that a portion of each print was scarified, but more than enough was left to match to his file prints.[19] Indeed, such scar patterns only call attention to themselves, and besides—like everything in nature—no two scar patterns are exactly alike!

Today, the FBI's Identification Division is its largest unit, with over twenty-six hundred employees. Files bulging with more than 200 million sets of prints represent more than 68 million individuals on file in the Criminal Justice Information Services Division. (Noncriminal fingerprint files are kept elsewhere.) All of this proves the accuracy of foresight of Francis Galton, who wrote in 1892: "Let no one despise the ridges on account of their smallness, for they are in some respects the most important of all anthropological data . . . . They have the unique merit of retaining all their peculiarities unchanged throughout life, and afford in consequence an incomparably *surer* criterion of identity than any other bodily feature."[20]

CLASSIFICATION

In order to be useful, fingerprints must be filed in great quantity, and searching for a match must be accomplished as quickly as possible. Alphonse Bertillon learned this lesson painfully. Although he had resisted fingerprinting in favor of his system of bertillonage and only reluctantly added fingerprints of a subject's right hand to his cards, he distinguished himself in 1902 as the first person in Europe to solve a murder with fingerprint evidence. (Edward Henry's earlier successes solving murders with his fingerprint system took place in India.)[21] Bertillon met with failure, however, in trying to solve the sensational theft of the *Mona Lisa*, stolen from the Louvre on August 21, 1911. The thief had left a clear thumbprint on the glass that had covered the priceless painting, but Bertillon had no system of classification for his thousands of cards with fingerprints. He and his assistants reportedly searched in vain through the files for several months. Then when the thief, Vicenzo Perugia, was arrested in Florence more than two years later, Bertillon learned that he had Perugia's fingerprints after all—those of his right hand—but the print at the scene had been from Perugia's left thumb![22] Brian Marriner wrote of the ironic moment: "It exposed the basic flaw in the system of bertillonage in brutal fashion, and Bertillon died on 13 February 1914 in the knowledge that his work had been in vain. He had been a scientific pioneer, but a failed one."[23]

In contrast to Bertillon, Sir Edward Henry lived to see his system of fingerprint classification become adopted and succeed. (After his service as Commissioner of Police in Bengal, he joined London's Metropolitan Police—Scotland Yard—in 1901 as assistant commissioner and became commissioner in 1903, serving until 1919.) With some modification, the Henry system is that in use today.

The system divides fingerprints into three basic pattern types: arches, loops, and whorls. (A fourth type listed in some books—composites, also known as "accidentals"—is considered a subtype of whorls: the accidental whorl.) These three main types are subdivided into eight distinct patterns.

ARCH	LOOP	WHORL
Plain	Radial	Plain
Tented	Ulnar	Central pocket loop
		Double loop
		Accidental

All human fingertips have friction ridges, and these ridges form the basis of all fingerprint patterns. As we discuss each of these pattern types in turn, keep in mind that the fingerprint expert utilizes detailed and complex sets of definitions to determine actual pattern types. This is necessary because sometimes a print has unusual features or looks deceptively like one type but technically is another.[24]

Arch patterns. The *plain arch* is one in which the ridges flow from one side of the pattern to the other with a rise or wave in the center. It has a smooth rise and a gentle upward curve. It is the simplest of all fingerprint patterns, and it generally gives no difficulty being correctly identified.

The *tented arch*, while a variety of the arch family, is somewhat more complex than the plain arch in its ridge formation. In its typical occurrence, it has either a central upthrust or a well-defined angle. It is regarded as a transitional pattern between arch and loop, and an example

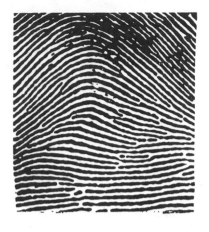

FIGURE 5.1. Plain arch. (Fingerprint images in figures 5.1 through 5.8 courtesy of FBI.)

FIGURE 5.2. Tented arch

FIGURE 5.3. Loop

FIGURE 5.4. Loop

may occasionally be seen with a "recurving" ridge or other feature common to the loop pattern; however, it does not have all four of the requisites of that pattern.

Arches represent only about 5 percent of all pattern types.[25]

Loop patterns. Both types of loop patterns have the same characteristics except for direction of flow. Basically, a loop has one or more ridges that enter on one side, recurve, and flow back out on the same side. To be a

FIGURE 5.5. Central pocket loop whorl

FIGURE 5.6. Plain whorl

FIGURE 5.7. Double loop whorl

FIGURE 5.8. Accidental whorl

true loop, the pattern must have four characteristics: a *core*, the approximate center of the pattern; a *delta*, or triangular area (analogous to the delta at the mouth of a river) caused by the divergence of ridges; at least one *recurving ridge* that passes *between* the core and the delta; and a *ridge count* of at least one. The ridge count refers to the number of ridges that cross, or at least touch, an imaginary line between the precise core and the precise delta. (For this purpose, fingerprint classification technicians have a fine line imprinted on a transparent device or *reticule*, that usually is inserted in the base of their fingerprint magnifier.)

The two types of loops, *radial loop* and *ulnar loop*, are based on the direction of flow of the ridges relative to the two long bones of the forearm: the radius (on the thumb side) and the ulna (on the little finger side). Therefore, to characterize a fingerprint as a radial or ulnar loop, one must know from which hand it came. On a right-handed print, if the ridges flow downward to the left, it is a *radial* loop; if they flow downward to the right, it is an *ulnar* loop. On the left hand the directionality is reversed, so that if the ridges flow downward to the left it is an ulnar loop; if to the right a radial loop.

Loops are the most common fingerprint pattern, appearing in about 60 percent of all fingerprints.[26]

Whorl patterns. A *plain whorl* pattern has two deltas and at least one ridge that makes a complete circuit about the core. Some have concentric circles (like a bull's-eye target) or ovals, others a spiral (like a clock spring).

The *central pocket loop whorl*, a complex pattern, is perhaps the most troublesome to classify. Essentially it is a combination of loop and whorl—that is, it has all the characteristics of a loop, with the addition of a second delta near the core and a whorl-type ridge or ridges circuiting around the core. One way to differentiate from the plain whorl is by drawing a line between the deltas: In a plain whorl such a line must cross at least one whorl-type ridge; in a central pocket loop whorl it must not do so.

The *double loop whorl* consists of two separate loop formations, each with its own core, and two deltas. (Sir Edward Henry did not designate this pattern. Instead, he recognized two double-looping patterns that he distinguished as follows: If the two loops exited on the same side of the print, they were termed *lateral pocket loops*, but if the loops exited on different sides they were called *twin loops*. The FBI learned that the distinction was more trouble to make than it was worth and consequently lumped the two patterns together into the category double loops.

Finally, the *accidental whorl* is a pattern with two or more deltas that may not be placed by definition into any of the other classes. For example, the pattern shown in figure 5.8 combines a loop and a whorl; but since it is not a central pocket loop whorl, it must be classified as an accidental whorl.

Whorls represent approximately 35 percent of all fingerprint patterns.[27] Fingerprints intended for classification and filing are recorded on standard eight- by eight-inch cards. These have spaces for *rolled impressions* from each finger (inked and rolled from one side to the other to record the entire pattern) plus spaces for *plain impressions* (the four fingers of each hand are pressed simultaneously, followed by the unrolled thumb print), which serve as checks on the rolled impressions to make sure they are recorded in correct sequence. At the top of the card the subject's name and fingerprint classification are recorded. On the reverse is recorded the *portrait parle* ("word picture") of the subject: height, weight, color of eyes, color of hair, scars, etc. (left over from Bertillon's system) and a place for full-face and profile photos (also a carryover from Bertillon, who invented the practice of taking "mug" photos[28]). Space for a "criminal history" is usually included on cards recording fingerprints of felons.[29]

After the prints are *recorded* and the information provided on the card, the first stage of the classification begins. This is called *blocking out* the set of prints, and it consists of placing in a space below each rolled impression a letter or symbol corresponding to its pattern type. (*All* whorls are marked W and ulnar loops are signified by \ for the right hand and / for the left. For the index fingers, the designations are A for plain arch, T for tented arch, and R for radial loop; for all other fingers the same symbols are used but in lowercase a, t, and r.) Next the ridge counts of all loop patterns are written in the upper right corners of their blocks. Finally, each whorl-type pattern is given a *ridge trace*. (This is a tracing of the ridge from the lower left side of the left delta to the right delta, dropping to a lower ridge anytime the ridge ends or forks. If the trace ends within one or two ridges of the right delta, a *meeting whorl* is indicated; otherwise it is designated an *inner whorl* or *outer whorl* depending on whether the traced ridge passes inside or outside the right delta.) The symbol of the appropriate ridge tracing—I, M, or O—is written in the upper right-hand corner of each finger block containing a whorl-type pattern to complete the blocking-out process.

The formula used for classifying the fingerprints may be composed of seven possible divisions: primary, secondary, subsecondary, major divisions, second subsecondary, final, and key. Their positions in the completed classification line are as shown below.[30]

Key	Major	Primary	Secondary	Second subsecondary classification	Final
	Divisions	Classification	Classification	Subsecondary Classification	

Note that some of the elements extend above and below the line. The second subsecondary classification—as its position indicates—is an

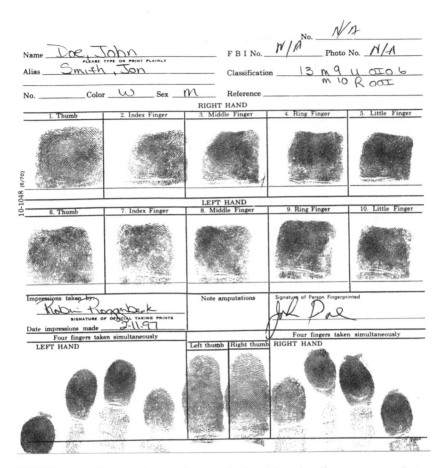

FIGURE 5.9. A fingerprint card records the ridge detail present on the first joint of the fingers, which are inked and fully rolled from side to side. These prints are used for comparison with those on other cards or with latent prints recovered from the crime scene. (Courtesy of Robin Roggenbeck, Orange County Sheriff's Office, Orlando, Florida.)

optional one, used when a group of fingerprints becomes so large that it needs further subdivision.)[31]

The classification process is exceedingly complicated, but the following cursory explanation of an actual classification—that of "Scarface" Al Capone—will provide an idea of the process. The *key* is the ridge count of the first loop—in Capone's case the right thumb, which has a count of 29. For the major divisions, the ridge counts of the thumbs are used; anything over 22 ridges is termed large, so Capone gets L/L for this part of the classification. For the next portion, the *primary classification*, whorls are recorded (using a formula in which different numerical values are assigned to each whorl, depending on which finger it appears on); however, since Capone's prints are devoid of any whorl-type patterns, he is assigned a simple 1/1 designation for his primary classification. The *secondary classification* represents the patterns of the index fingers, so Capone's ulnar loops translate as U/U. Next, the subsecondary classification is based on the three middle fingers of each hand with whorls being designated I, M, or O and loops (depending on their small or large ridge counts) being represented as I or O. Here Capone's loop counts for each hand yield an IOI/IOI subsecondary classification. The *final classification* records the ridge count of the right little finger (with provisions for that print being a pattern other than a loop). In the case of Capone's little finger, the count is 6, so that number is entered at the extreme right, above the line. The completed classification is therefore:

29	L	1	U	IOI	6
	L	1	U	IOI	

The second subsecondary classification is omitted.[32]

Such a classification system permits fingerprints to be filed and subsequently retrieved. The primary classification alone permits fingerprint cards to be divided into 1,024 groups; however, the 1/1 primary classification of Capone's prints would include about 25 percent of people, so the need for the other subdivisions can easily be appreciated.

Criminals have never been considerate enough to leave rolled impressions of all ten fingers at the scenes of their crimes. Although TV detectives always seem capable of identifying the culprit who has left a single latent thumbprint, for real law enforcement personnel it is not such an easy matter. Historically, the latter had usually been limited to searching the files of known suspects. In 1933, the FBI did address this longtime problem by inaugurating a single-fingerprint file, using modifications of a method developed by former Scotland Yard Chief Inspector Harry Battley. The method utilized a special reticule, half of which had a series of concentric circles that were used to define specific areas of scrutiny. Also, some pattern types were further subdivided (arches into five categories, for instance), and ridge counts for whorls were added. But such additional work multiplied tenfold the work of classificating fingerprints, so—as a practical matter—files were kept only for certain notorious criminals, repeat felons, and those in certain categories, such as bank robbers and kidnappers.[33]

Since 1970, however, computer technology has made it possible to automate the process of fingerprint searching. Forensic expert Richard Saferstein describes these automated fingerprint identification systems (AFIS):

> The heart of AFIS technology is the ability of a computer to scan and digitally encode fingerprints so that they can be subject to high-speed computer processing. The AFIS uses automatic scanning devices that convert the image of a fingerprint into digital minutiae [ridge characteristics] that contain data showing ridges at their points of termination (ridge endings) and the branching of ridges into two ridges (bifurcations). The relative position and orientation of the minutiae are also determined, allowing the computer to store each fingerprint in the form of a digitally recorded geometric pattern. The computer's search algorithm determines the degree of correlation between the location and relationship of the minutiae for both the search and file prints.[34]

He adds:

> In this manner, a computer can make thousands of fingerprint comparisons in a second; for example, a set of ten fingerprints can be searched against a file of 500,000 ten-finger prints in about eight-tenths of a second. During the search for a match, the computer uses a scoring system that assigns prints to each of the criteria set by an operator. When the search is complete, the computer then produces a list of file prints that have the closest correlation to the search prints. All of the selected prints are then examined by a trained fingerprint expert, who will make the final verification on the print's identity. Thus, the AFIS makes no final decisions on the identity of a fingerprint, leaving this function to the eyes of a trained examiner.[35]

Now the AFIS processing makes it possible to search a single crime-scene latent print against an entire collection of fingerprints files. In one case, a serial killer who had terrorized Los Angeles with fifteen murders was identified in about twenty minutes after AFIS was brought into service; it was estimated that manual searching would have required a technician, searching manually through that city's 1.7 million fingerprint cards, some sixty-seven years to accomplish the same results.[36]

IDENTIFICATION

Once fingerprints have been recorded, classified, and filed, they always can be compared to others, whether latent prints recovered from a present

FIGURE 5.10. An Automated Fingerprint Identification System (AFIS) is used to compare one inked fingerprint with another to establish an individual's identity. (Courtesy of Robert Ruttman, Mesa Police Department, Mesa, Arizona.)

crime scene or another set of prints taken and classified by a different technician years later. The method of comparison is to use the ridge characteristics or minutiae to *individualize* the print, establishing the individuality or uniqueness of it. Many fingerprint examiners still use the term *identify*, although the word is imprecise. Tradition aside, one should actually speak of *identifying* a marking as a fingerprint but of *individualizing* a fingerprint as that belonging to a particular person.)[37]

The ridge characteristics or minutiae used to individualize a fingerprint include those utilized in computerized AFIS technology: any *ridge ending* (termination point of a friction ridge) and what is known as a *bifurcation* (a branching or forking of a ridge into two ridges). There are two additional common ridge characteristics, both recognized by Sir Francis Galton and sometimes termed in the past "Galton Details": *ridge dot* (a ridge feature that resembles a period—i.e., it is only about as long as it is wide) and *enclosure* (a ridge characteristic resembling an eyelet, caused by the legs of a bifurcation coming together again to form a single ridge).[38] From time to time certain other terms are employed, including "short ridge" (one whose terminal points are very close together), "trifurcation" (a branching into three bridges), and a "bridge" (or bar linking two ridges).[39] Creases, which often show in fingerprints as short white lines crossing the ridges, are not considered *minutiae* because they are impermanent features. A scar, however, may be utilized as such for purposes of comparing two prints. (A scar may typically be distinguished from a crease because the former results in a puckering of the ridges along its length.)[40]

To actually make a comparison, the expert looks for four elements; if the elements of one print match those of another, individualization may be declared.

1. *Likeness of pattern.* Two prints must have a likeness of pattern types (arches, loops, etc.).

2. *Qualitative likeness of ridge characteristics or minutiae.* The ridge endings (enclosures, bifurcations, etc.) must match.

3. *Quantitative likeness of ridge characteristics.* A sufficient number of ridge details must be present for individualization to be declared.

4. *Likeness of location of minutiae.* The friction ridge details must be in the same relative positions in both fingerprints—that is, they must be the same relative directions and distances from each other, with the same number of ridges between them in both fingerprints.[41]

Also, there must be no unexplained differences between prints.[42] As to the number of ridge characteristics or minutiae that must be the same for a "match" to be declared between two prints, the issue has been much debated; numbers as low as eight and as great as sixteen have been proposed.[43] Ten to twelve characteristics would generally seem a reasonable number in most cases, fewer if the points of comparison are particularly distinctive.[44] Sometimes, when very few ridge characteristics are present but the *pores* in the ridges are visible, the science of *poroscopy* may be applied. As first noted by Edmund Locard, the shape, size, and position of the pores are also different from person to person, are permanent, and may serve as the basis for a positive individualization of one print with another.[45]

To facilitate making a comparison, the fingerprint expert may use an imaging device such as a "comparator," which places the two images—greatly enlarged—side by side on an illuminated viewing screen. Modern electronic imaging technology, which uses digital photography (and thus eliminates processing of photographic film), can also provide nearly instant enlargements of fingerprints for comparative purposes. Such

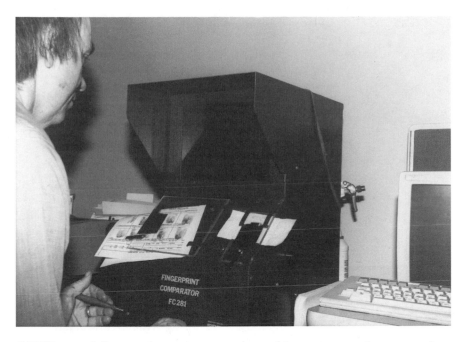

FIGURE 5.11. A fingerprint comparator is used by some examiners to make a comparison between a known inked fingerprint from a standard ten-print card and a questioned latent print. (Courtesy of Tony Moss, Orange County Sheriff's Office, Orlando, Florida.)

images can even be printed on acetate for display by an overhead projector, or may be printed out on paper for exhibit purposes.[46]

For courtroom presentation, the known file print and the questioned latent print are compared side by side in greatly enlarged form. The standard method of demonstrating similarity is to draw a fine line from each ridge characteristic to the white space surrounding the print, and to give each such point its own number. To avoid confusion, placement of lines and numbers should be as similar as possible on both prints. Usually, the numbers and the names of the features they represent are listed on the exhibit.[47]

In trying to circumvent the conclusive evidence of fingerprints, criminals sometimes outdo themselves. A case in point concerned a burglary committed in Israel in 1968. Although no fingerprints were evident on the window that had permitted entry or elsewhere at the scene, bare footprints were recovered. When a suspect was arrested a month later, he seemed surprised at the evidence against him. "Before I broke into the place," he explained, "I took off my shoes and put socks over my

FIGURE 5.12. A fingerprint examiner uses a magnifier, usually one specially designed for the purpose, to determine if a fingerprint is of value and to make subsequent comparisons. (Courtesy of Robin Roggenbeck, Orange County Sheriff's Office, Orlando, Florida.)

hands so I wouldn't leave any fingerprints." This lack of knowledge about the uniqueness of footprint evidence cost the man two years in prison.[48]

In fact, any area of the inner surface of the hands or feet contains friction ridges and patterns that are just as effective for comparison as are fingerprints. The only drawback is the difficulty of recording and classifying such impressions. However, when such evidence presents itself, criminalists know its evidentiary value and are quick to see that it is preserved for possible later comparison. A case illustrating this took place in England in 1955 when a woman was murdered on a golf course. The murder weapon, an iron tee marker lying near the body, had a bloody palmprint on its surface. Since palmprints were not part of police files, investigators determined to record palmprints from people who lived in the area. Nine thousand impressions were taken, and at just over the halfway point—the 4,604th palm imprint to be examined—a match was made between the questioned print on the murder weapon and the

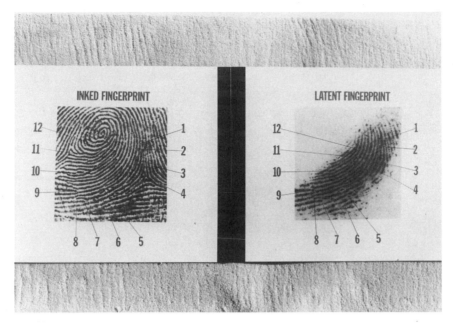

FIGURE 5.13. A fingerprint chart is used in court to explain to jurors how a latent fingerprint is compared to a recorded print and identified. The chart illustrates some of the characteristics that match in the two prints. (Courtesy of Tony Moss, Orange County Sheriff's Office, Orlando, Florida.)

palmprint of a seventeen-year-old government employee. He pleaded guilty and was sentenced to prison.[49]

PROCESSING AND RECOVERY

Fingerprints may be recovered from a crime scene in several forms. It is usual to list three types of prints discussed presently, but we will add a fourth, special type that might be considered as part of the *gruesome* category. In one instance, in Cochise County, Arizona, in 1919, at the scene of a burgled safe, a criminal left behind a crowbar and other wrecking tools, plus some splotches of blood and a piece of skin from a finger— complete with ridge patterns! He was thus identified and sentenced to five to fifteen years in the Arizona State Penitentiary.[50]

Earlier, Scotland Yard had recorded an even more gruesome case. Early one morning a sharp-eyed constable spied an odd-looking object at the top of an iron spike on a warehouse gate. It proved to be the right ring finger—complete with a cheap ring—that a would-be thief had left behind. Obviously he had caught his ring on the spike, and his own weight had caused the finger to be torn off. A short time later at the Yard's Fingerprint Bureau, the finger's ridge pattern was matched to a file print, that of a burglar with a lengthy criminal record.[51]

Still another example, perhaps the most gruesome of all, comes from a foiled robbery of the Farmer's National Bank in Pittsburgh in 1926. The robber's note said that he had a bomb, but the teller was skeptical and set off the alarm. True to his word, the would-be robber detonated his parcel, killing two, and blowing one—himself—to bits. Among the bits were an ear and a right hand. The latter was inked and printed, run through FBI files, and promptly identified as that of one William Chowick, a repeat criminal.[52]

These cases are relatively rare. Much more common are the three types of finger impressions that may be found at the scene of a crime. *Plastic fingerprints* are those impressed into some substance such as wax, soap, putty, or even dust. They are three-dimensional depressions made by the friction ridges. *Visible prints* are those left by fingers that have been coated with some colored substance such as blood, grease, paint, dirt, or ink. *Latents* are fingerprints that (as the name implies) "lie hidden" or are relatively invisible and need some form of processing or developing.[53]

Plastic and visible fingerprints may easily be photographed, but latent prints require careful handling. To see the supposed development of latents on some detective shows on television, one would think the technician was attempting to scrub the stubborn spots away, but Charles E. O'Hara describes their "delicate nature":[54]

> To deposit a thin layer of perspiration or grease in the complicated pattern of the friction ridges optimum conditions must be present. The surface must be such that it can retain the print without absorbing and spreading it. Thus hard, glossy objects such as glass and enamel painted walls and doors present ideal surfaces. Dirty surfaces and absorbent materials do not readily bear prints. The fingerprint, moreover, must be deposited with the right amount of pressure. The object must not be touched with an excess of pressure, since this tends to spread the print. A movement of translation of [*sic*] the finger will result in a smear. The fingers of the person depositing the prints must have a certain degree of moisture or should have some body grease on the ridges. When all these requirement are fulfilled a good latent fingerprint is deposited. Despite the infrequency with which a latent fingerprint examination meets with success, the unsurpassed value of a print as evidence warrants the expenditure of effort that this search entails.[55]

The thin film that comprises a latent fingerprint contains certain substances, predominantly perspiration, secreted by the pores in the friction ridges, and/or, body oils, acquired by contact with other parts of the body, notably the hair and face. The perspiration is composed primarily of moisture with the added presence of sodium chloride (salt), amino acids, and generally to a lesser degree, a variety of other organic and inorganic substances. There also may be traces of other materials such as dead epidermal cells and various foreign substances. The decision regarding what treatment should be used to develop a latent print depends largely on the surface on which it lies, experience with the substances in such prints, and the technology available (small law-enforcement agencies may not have access to the most advanced technology).[56]

At the scene of a crime, a search is conducted for fingerprints by specially trained personnel. The FBI advises proceeding in a logical fashion, with care given to points of entrance and exit, for example, and any objects or surfaces that might have been disturbed or touched during the criminal act. Wearing light cloth gloves while collecting evidence is recommended, although some experts use disposable latex gloves, and one authoritative source recommends against gloves entirely,[57] since they invite carelessness that may destroy prints or leave glove prints that mislead others. Objects should be handled minimally and then only by the edges or by surfaces that are unreceptive to fingerprints, such as the knurled handles of pistols. The fingerprint expert should be given priority access to objects such as firearms and documents over other criminalists such as the firearms examiner or questioned-document expert. As we have emphasized before, any articles removed from the scene must be correctly labeled and carefully transported to prevent destruction of prints.[58]

In actually looking for fingerprints, as for other trace evidence, a flashlight and magnifying glass represent the standard equipment. The beam of the flashlight played at an acute angle to the surface may help reveal latent impressions. Examining the surface from various angles may accomplish the same purpose.[59] When a plastic or visible fingerprint is discovered, it should be photographed *in situ,* as should a latent impression after it is developed.

A standard approach for latent fingerprints discovered on nonporous surfaces is to begin with a visual search and then extend that, if necessary, to a search with more sophisticated light sources such as the argon laser beam, which causes the latent print residue to *fluoresce.*[60] B.E. Dalrymple and Roland Menzel played critical roles in the application of

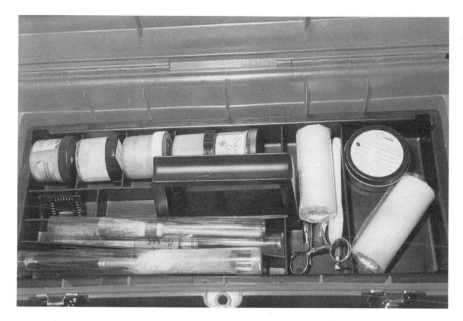

FIGURE 5.14. A typical fingerprint kit with its variety of latent brushes and developing powders. (Courtesy of Ed Hobson, Orlando Police Department, Orlando, Florida.)

the laser to fingerprint detection.[61] Since April 1978, the FBI Latent Fingerprint Section has used laser technology to detect latent impressions on various types of evidence,[62] and the procedure is now standard in most crime laboratories. According to the FBI, "The procedure used to detect latent prints with a laser is clean, relatively easy and no pretreatment of the specimens is required. There is no alteration of the evidence; therefore it is used in most instances prior to the application of conventional processes."[63] In addition to lasers, there are certain high-intensity light sources, or "forensic light sources," that have all but replaced the laser. There also is the Reflected Ultraviolet Imaging System (RUVIS), which makes use of wavelengths beyond those of visible light, specifically those in the ultraviolet (UV) region. The UV rays can reveal latent finger impressions that might otherwise go undetected.[64]

Following laser detection of a fingerprint on a nonporous surface, the next step for relatively small objects is to develop the print using cyanoacrylate vapor, a substance that is known to the public as superglue. The technique was brought from Japan to the United States by Ed German and Paul Norkus of the U.S. Army crime lab. Frank Kendall improved the technique in 1982 by using the glue in conjunction with

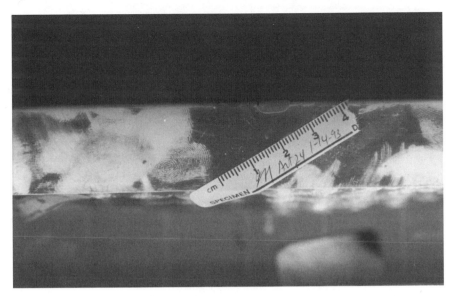

FIGURE 5.15. One of the most commonly used latent-print processing techniques for nonporous objects is superglue or cyanoacrylate fuming. Shown here is an "ammo" box that has been fumed, resulting in the development of several latents.

sodium hydroxide. The fuming process is carried out in an airtight tank. Polymerization occurs when vapor adhering to the latent residue during condensation builds up and hardens the ridges.[65] Following the superglue fuming, treatment of the evidence with dyes like rhodamine 6G causes the impression to fluoresce under laser illumination, which may increase detection.[66]

Instead of this fuming technique, or in addition to it, the fingerprint expert may use a brush to apply fingerprint powders. The standard ones are lampblack for use on light-colored surfaces and titanium oxide for dark surfaces. Occasionally, colored powders, such as dragon's blood, are used for certain special situations. Alternatively, a magnetic-sensitive powder is applied with a magnet called Magna Brush. Since there are no brush bristles with this method, there is less chance the print will be damaged. There are also special fluorescent powders for use with the laser or other forensic light sources. The reason for using first the forensic light sources, then the chemical fuming, and finally the dusting with powders is to be as nondestructive as possible. Powder may result in smudging or overprocessing of the print, and once it is used, other procedures are generally precluded. After a print is developed by powdering,

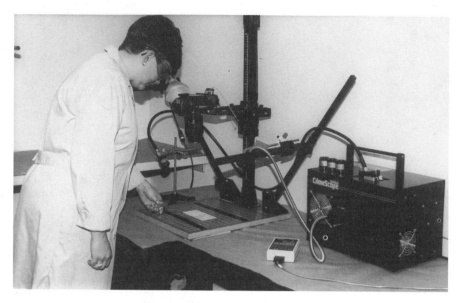

FIGURE 5.16. A typical visible-range forensic light source, such as the "CrimeScope" pictured here, can provide almost any color of light that may best develop or enhance a latent print. The print is then photographed using a close-up lens and special filters.

it may be lifted with special transparent tape and then protected with a stiff, transparent cover. Since the lifting process is somewhat risky and may damage or distort the print, the impression should be photographed before lifting tape is applied.

Porous surfaces such as paper, cardboard, and unpainted wood present more difficulty. With such surfaces, powders should be avoided since they tend to adhere to the background surface as well as the print. Following visual and fluorescent examination, one may use instead chemical treatment with ninhydrin—which reacts with the amino acids in the perspiration residue—to develop any latent impressions that might be present. The ninhydrin solution is usually sprayed on, and there are aerosol cans made specially for the purpose.[67] Instead of ninhydrin treatment, the examiner may use a substitute called DFO (1,8-diazafluoren-9-one). This "visualizes" (as criminalists say, meaning "renders visible") latent fingerprints when they are exposed to a forensic light source, which causes them to luminesce.[68]

Other techniques used for porous materials include application of a solution of silver nitrate (which reacts with the sodium chloride in the perspiration residue)[69] or of another silver-based solution called "physical developer" (which reacts with oils and fats). For developing prints on wet surfaces, a material called "small-particle reagent" may be used.[70]

One of the most difficult surfaces to develop fingerprints on—yet one of the most useful as probative evidence—is human skin. The fact that murderers frequently grab their victims makes it desirable for homicide investigators to develop identifiable fingerprints from the skin of a corpse. Unfortunately, one of many problems in accomplishing this is that the victim's skin contains the same substances, such as perspiration and oils, found in fingerprint residue, thereby making latent print development quite difficult. In 1991, a joint research project of the FBI in cooperation with the Knoxville Police Department, the Department of Anthropology at the University of Tennessee, and the University of Tennessee Hospital addressed this problem.

Eventually this project found employing the cyanoacrylate fuming technique and then applying magnetic fingerprint powder to be the most effective procedure. Application of this technique to a corpse, however, presented problems, and attempts to confine the fumes by using a plastic tent over a portion of the body were not very successful. At length, the researchers developed a fuming device that heated the superglue in an aluminum pan and dispensed the fumes through a plastic hose. Researchers also found that the kind of powder used was less significant than correct application of the cyanoacrylate fumes, which develop prints in five to fifteen seconds, depending on body temperature. Because of condensation on refrigerated bodies, it is recommended that, if possible, prints be recovered before the body is refrigerated; otherwise, the process should be delayed while any moisture evaporates.[71]

Even fingerprints that have not developed well often may be improved by use of a number of sophisticated instruments. These include such readily available computer software programs as Adobe Photoshop, which can enhance a latent print by improving the image's sharpness and contrast.[72] Even the common Kodak CopyPrint Station—which is found in many photographic shops where customers make copies or enlargements of their photos—can produce photographic quality fingerprint enlargements in minutes.[73] Also there are filtering techniques, such as an Iterative Automated Noise Filtering method, that can remove electronic "noise" from digitized fingerprint images, thereby enhancing ridge detail.[74]

Modern law enforcement continues to experiment and to apply the latest technology to solve crimes—and nowhere more diligently than in the crucial area of fingerprinting. Because of scientific advances, criminals who leave latent prints at crime scenes, and even on the skin of

victims, "will just be putting themselves"—according to one FBI expert—
"within easy reach of the long arm of the law."[75] The continuing value of
the science of fingerprints is well illustrated in the following case study.

CASE STUDY: AILEEN WUORNOS, FEMALE SERIAL KILLER

> One guy wasn't dead after I shot him. He just fell down and was sayin':
> "I'm gonna die. Oh, my God! I'm dying." I said: "Yeah, motherf——r,
> so what?" And I shot him a couple a more times.
>
> —Aileen Wuornos[76]

The first body was found by "three scavengers," as crime writer Michael
Reynolds describes them in his book *Dead Ends*. Two of the scavengers
were Florida men scrounging through a wooded dumpsite northwest of
Daytona looking for scrap metal. Says Reynolds of the third: "The buzzard was there to dine."[77] One of the men described finding the body in
the semiliterate, handwritten report he later produced for investigators
from the Volusia County Sheriff's Office: "I James A Bonchi & James
Davis were out skraping mettle's [metals] and exploring throw [through]
the woods and seen a Buzzard and, trompeding [tromping] through the
woods we smelled stink! I was off troming [roaming?] through the Palms
when my friend yelled for me and sed look! I saw a tarp [tarpaulin] with
a hand hanging out of It!"[78]

Twelve days earlier, an abandoned Cadillac had been discovered north
of Daytona. It was traced to a Richard C. Mallory, age fifty-one, the missing owner of an electronics repair shop in Clearwater.[79] When investigators arrived at the rural site discovered by the scavengers, they quickly
marked off the area with yellow crime-scene tape and—without touching the still-covered body—began to search for evidence. In the meantime, as darkness approached, a crime-scene specialist for the department's
Major Case Unit took photographs—both Polaroid and 35mm—and used
a tape recorder for the necessary documentation. When the van arrived
from the Florida Department of Law Enforcement's (FDLE) regional
crime lab, the criminalists turned on the van's powerful spotlights and
began taking their own photos and measurements. Then, donning surgical gloves, they carefully lifted the "tarp," actually a filthy piece of red
carpet runner, from the corpse. A cloud of steam—vapors from the decomposing body—rose into the cool air with a hissing noise. Two pieces
of cardboard also were lifted off.

The body was lying face down. From the collarbone up there was only
black decay, swarming with maggots and other insects. After investiga-

tors photographed and videotaped the body, they turned it over onto the back, whereupon the lower jaw dislodged, spilling out a set of dentures. Pointing to the chest, sheriff's office investigator Larry Horzepa said: "He bled from there. And up. The insects follow the blood." The deceased man's clothing was intact except that his right loafer was off and his pockets had been turned out. Everything, including the clothing, was consistent with the corpse being that of the missing Richard Mallory. The criminalists gathered items of evidence, including scraps of paper and a sample of hair near the body, and placed them in brown paper bags. After the FDLE criminalists left, the investigators waited until the hearse came and then helped place the corpse into a body bag and load it aboard.[80]

The next day an autopsy was conducted at the office of the Daytona Beach medical examiner. One bullet had fallen from the body and was recovered from the body bag; three more were located by x-ray and recovered by opening the chest. More difficult was obtaining a set of fingerprints. In the case of the newly dead, this is usually accomplished in the standard fashion, by inking the prints, except that—since it may be difficult to roll the fingers on the card—the opposite is done: after the fingers have been inked, the individual print squares are cut from the card and each is rolled around the finger, usually by means of a spoon-shaped tool that holds the squares (or usually strips of five squares).[81] In obtaining fingerprints from decomposed bodies, the fingers may be injected with water or glycerin using a hypodermic syringe if the skin is wrinkled (as it may be with drowning victims). In more advanced cases of decomposition, any one of a number of procedures may be used, including photographing the prints or thinly coating them with a heavy salt such as bismuth or lead carbonate and then x-raying them. Sometimes the skin is cut from the fingers so that it can be flattened before either photographing or inking and printing the ridge patterns. Occasionally the skin of the hand slips off like a glove and is treated as such: the examiner inserts his own latex-glove-covered hand into the "glove" and rolls it on the inking pad and then on the fingerprint card.[82]

When the necessary fingerprint technology is not present at the time of autopsy, the fingertips or the entire hand may be amputated for transport to the crime laboratory. The hands, fingers, or skin so transported are typically placed in embalming fluid or a 5-percent solution of either formaldehyde or alcohol. If the tissue is desiccated (dried out), it may be sealed in airtight bags and placed in an ordinary cooler. This was the case with the Volusia County murder victim, whose entire hands were transported to Orlando by crime-scene specialist James Lau. At the FDLE laboratory, the hands were removed from their bags and placed in a strong saline solution for about two hours in order to tighten the skin. In the meantime, fingerprints of Richard Mallory, taken when he had been arrested for drunken driving, were located in the files in Tampa and were faxed to the FDLE lab for comparison. The saline treatment permitted prints to be obtained from one hand, and these were compared to the file prints. They matched, thus identifying the body of the murder victim as Richard Mallory.[83]

Criminalist Lau phoned the news to Horzepa, then cleaned the ink from the fingers and rebagged the hands for transport back to the medical examiner's office. There the hands would be reattached to the body so that it could be released to a funeral home. The reattachment needed to be done as soon as possible, so Lau sped on his way. Occasionally, even in a serious homicide investigation, there may be a lighter moment or two, and this was no exception, as Michael Reynolds relates in *Dead Ends*:

> In an unmarked car, Lau sped up Interstate 4 and into the radar field of a Florida Highway Patrol trooper, who immediately pulled the investigator over.
>
> Lau handed over his license and registration to the trooper, who then asked, "What's the hurry?" "I've got these hands here. Got to get them back."
>
> "What?"
>
> "Here," the ever-friendly Lau offered, picking up the cooler from beside him and opening it in the trooper's face. "These hands here."
>
> The stench flew up to the trooper's face.
>
> "Good god!" The trooper took a quick step backward.
>
> "Jeez, you guys . . . get them out of here. Go on."
>
> "Okay. Thank you, trooper." Lau smiled and shut the box of hands. He drove the rest of the way back to Operations just as fast as he thought necessary. If another FHP wanted to take a peek, he was welcome to it.[84]

It was six months before another victim was discovered, followed by still another five days later, then another and another—seven in all. All were men, and all had been murdered along central Florida highways:

1. As described, Richard Mallory was the first victim. He had been shot four times with a .22-caliber weapon. His body was discovered on December 13, 1989, in Volusia County. Mallory's car had been discovered twelve days earlier, along with his empty wallet, a half bottle of vodka, and some condoms.

2. On June 1, 1990, the body of David Spears, age forty-three, a construction worker from Winter Garden, Florida, was discovered in a desolate area of Citrus County. Found naked except for a baseball cap, he had been shot six times with a .22. He was identified by means of dental x-rays.

3. On June 6, the nude body of Charles "Chuck" Carskaddon, age forty, was found in Pasco County. A part-time rodeo worker, Carskaddon had been shot six times with a .22-caliber firearm. He had been on a trip to Tampa to visit his girlfriend.

4. The following day, Peter Siems, age sixty-five, was listed as missing. A missionary who had formerly been a merchant seaman, Siems had set out to visit relatives in New Jersey and Arkansas. His car was found abandoned on July 4 in Marion County, but his body was never discovered.

5. Troy Burress, a forty-six-year-old deliveryman for the Gilchrist Sausage Company in Ocala, was the next murder victim. Missing since July 31, when his abandoned delivery truck was discovered about 4:00 A.M., Burress was found dead August 4 in the Ocala National Forest, in Marion County, by a family on an afternoon picnic. He had been shot in the chest and back. His wallet was missing, but his wife identified his wedding ring and provided the address of Burress's dentist; the dental information led to a more definitive identification.

6. A third Marion County victim was Charles "Dick" Humphreys, age fifty-six. A retired Alabama police chief, he had become a social services investigator in Sumterville. On September 11, 1990, the day after his thirty-fifth wedding anniversary, Humphreys disappeared. His bullet-riddled body was discovered the next day. His pockets had been pulled out, and his wallet was missing.

7. Walter Gino Antonio, a sixty-year-old Brevard County Reserve Deputy, was the last victim. His body, nude except for socks, was found November 19 on a remote logging road in Dixie County. He had been shot four times, once at the base of the skull.[85]

By the time of Humphreys's death, investigators from the various counties had begun to share information and had concluded that a serial killer was at work. One interim death, that of a man named Giddens, had occurred in Marion County, but he had been shot a single time with a .38. Also, the tags of the cars of all the other victims had been removed, and the cars had been wiped clean of fingerprints. Except for Giddens, all of the cases had similar elements. During November 1990, the various investigators involved had begun to hold strategy sessions at the Marion County sheriff's office. The prevailing theory was that the men were being solicited either for automotive assistance or for sex with the intent to murder and rob them.

In the meantime, detectives had turned up a strong lead. On July 4, two rough-looking females had been seen leaving the car that belonged to Peter Siems, and the witnesses gave information for sketches of the pair by police artists. One sketch depicted a woman with stringy blond hair, thin lips, and narrow eyes, while the other drawing showed her companion, a round-faced woman wearing a baseball cap.

On November 29, the multi-agency task force went public at its command post in the Marion County sheriff's office. That county's chief criminal investigator, Capt. Steve Binegar, briefed the assembled press and handed out copies of the police sketches. The story went national within twenty-four hours, appearing on the various network news programs and in *USA Today*. Plans were made to funnel any incoming leads into a computer.[86]

Leads were not long in coming. The day the news story appeared, a man from Homosassa Springs called to say that he recognized the two women in the sketch as Tyria Moore and another woman known as Lee. The pair had once rented a trailer from him. Within three weeks, a total of four leads had been logged—all naming the same women. An anonymous caller had given much information, including the names Lee Blahovec and Tyria "Ty" Moore, and claimed that the two were "lesbians, were violent, and had a hatred for men."[87]

A search through FDLE computers revealed driver's licenses and criminal records for both Moore and Blahovec, the latter having used the alias Cammie Marsh Greene. Horzepa and other Volusia County sheriff's officers began checking pawn shops in the Daytona area, looking for anything that might have belonged to one of the victims. On December 21, they got lucky. On December 6, two items of Richard Mallory's—a Microna Road Patrol radar detector and a 35mm Minolta Freedom camera—had been pawned at a shop on Second Avenue. The woman had given her name as Cammie Marsh Greene and, as required by law, *had placed her right thumbprint on the pawn receipt.* Investigator Bob Kelley rushed to Orlando to give the receipt to FDLE fingerprint experts.[88]

The thumbprinted receipt was received by Jennie Ahern, supervisor of the Latent Print Section of the FDLE's Orlando Regional Crime Laboratory. She and crime laboratory analyst Debbie Fischer (wife of co-author John F. Fischer) worked into the evening attempting to identify the single print using the computerized AFIS technology. Although the minutiae (ridge characteristics) of the thumbprint were of sufficient clarity and number for an individualization to be made, AFIS turned up no matching print. Ahern then called the Volusia County investigators and volunteered to visit the various law enforcement agencies and manually search their fingerprint records. Such files are known to contain records that, for various reasons, such as dropped charges, are not in AFIS or FBI files.

The following day, December 22, 1990, Jennie Ahern, Debbie Fischer, and another expert, David Perry, went to the Volusia County sheriff's department where they began a "cold search"—a card-by-card search, by hand, through a file. They first extracted cards with whorl-type prints, then split those into three groups and sat down to what they thought would be the first stage of a laborious task. As it happened, however, after only about fifteen minutes, or about a hundred cards each, Jennie Ahern got lucky and found the matching thumbprint, which was then confirmed by Fischer and Perry. It was on a card with the name Lori Kristine Grody.[89] She had been arrested on a weapons charge in 1986, having occupied a stolen vehicle with a loaded .22-caliber pistol hidden under the front seat and with extra rounds in her shoulder bag.[90]

Grody's photograph was obtained from jail records and was found to match that of Susan Blahovec. Once again, the evidence was pointing at the woman known as Lee. Investigators then sent Grody's prints to the FDLE lab in Tallahassee where they were matched to a bloody handprint that had been left in Peter Siems's Sunbird. At this point, the Grody record, along with the record of Ty Moore, was run through the National Crime Information Center (NCIC)—the computerized national clearinghouse for information on fingerprints and photographs.[91] This turned up responses from law enforcement agencies in Michigan, Colorado, and Florida, which finally identified "Lee"—alias "Lori Kristine Grody," alias "Cammie Marsh Greene," alias "Susan Lynn Blahovec"—as one Aileen Carol Wuornos, a white female born on February 29, 1956 (leap year), at Rochester, Michigan. She was described as five feet, two inches tall, with medium build, light brown hair, and brown eyes. According to writer Michael Reynolds, "Juvenile records for truancy, being a runaway, shoplifting, and burglary paled beside her adult exploits."[92] The charges included armed robbery (which had earned her just over a year in prison from 1982-83), grand theft auto, resisting arrest, disturbing the peace, assault and battery, attempting to pass forged checks, prohibited use of a firearm (firing a .22 pistol from a moving vehicle), and so on and on. A Florida Highway Patrol officer, who gave her a ticket for driving seventy-two miles per hour in a fifty-five-miles-per-hour zone, noted, "Attitude poor. Thinks she's above the law."[93]

Photographs of both Moore and Wuornos were distributed to other investigators, and tips soon came flowing in. Believing Wuornos to be in the Harbor Oaks vicinity, the task force set up a command post at a motel in nearby Daytona. In addition to a surveillance van and various cars, there were thirty officers from six agencies, including two undercover agents. The operation began January 5, 1991, and within two days, "Lee" Wuornos had been spotted at a bar and surveillance had begun. The undercover men were soon drinking with her; one even danced with her. After a while, she walked on to another bar, "The Last Resort," where she remained for the rest of the night. Except for two units—one at the front and one at the rear of the bar—the surveillance was discontinued so that everyone could get some sleep.

At two o'clock the following afternoon, the undercover officers entered "The Last Resort" wired for sound. They engaged Lee in conversation and eventually learned that her best friend (Ty Moore) had taken off and that she had no place to stay. Finally, the officers listening in on the conversation decided to arrest her when she left the bar. At about 6:00 P.M., she was arrested on the outstanding warrant for having a concealed weapon. She was unarmed at the time, and the .22 pistol was not found in her suitcase or bag, which were retrieved from the bar.

Tyria Moore, located in Pennsylvania, agreed to return to Florida and assist in the investigation. After many calls that she received from Wuornos in jail—calls that the police listened in on with Moore's permission—Moore eventually persuaded Wuornos to confess—ostensibly to keep her (Moore) from being charged for crimes she did not commit. Wuornos gave a lengthy account of the murders, although she maintained that she always killed in self-defense. She also told detectives where to find her pistol: in the waters of Rose Bay. Bullets test-fired from it matched those that had killed Richard Mallory.[94]

Although a spokesperson for the FBI stated that Aileen "Lee" Wuornos was "the first female textbook case of a serial killer,"[95] she was actually the thirty-fifth such murderer in U.S. criminal history. Explains Reynolds: "The difference with Lee was that she targeted adult male strangers as her victims and dispatched them with a firearm. Most female serial murderers chose to kill members of their families, persons in their care, or close acquaintances, and usually dispatched then with poison. Typically, they were 'quiet killers,' working within the confines of their home, a hospital, or an elderly-care facility. Lee was certainly not 'quiet' and searched out her victims, all strangers, along the highways of central

Florida."[96] Another writer styled her "America's first *sexual* serial murderess" (emphasis added).[97]

On Monday, January 14, 1992, Wuornos's trial for the murder of Richard Mallory began in Deland, Florida. It lasted just two weeks, with jury deliberations beginning on January 27. The jury took just one and a half hours to bring in a guilty verdict. The penalty phase began the next day, and on January 30, the same jury recommended death. The following day the judge sentenced her to die in the electric chair. She later stated that she would not contest the additional murder charges, and she pled no contest to three murders in Citrus and Marion Counties.[98]

NOTES

1. Gerald Posner, *Case Closed: Lee Harvey Oswald and the Assassination of JFK* (New York: Random House, 1993), 283.

2. Eugene Block, *Fingerprinting: Magic Weapon against Crime* (New York: David McKay Co., 1969), 3.

3. Ibid.; Martin Fido, *The Chronicle of Crime* (New York: Carroll & Graf, 1993), 253.

4. Richard Saferstein, *Criminalistics: An Introduction to Forensic Science,* 5th ed. (Englewood Cliffs, N.J.: Prentice Hall, 1995), 412.

5. Henry Morton Robinson, *Science Catches the Criminal* (New York: Blue Ribbon, 1935), 42-43; Block, *Fingerprinting,* 1-4; *A Study of Finger Prints: Their Uses and Classification,* 33rd ed., forensic instruction manual for course in Scientific Crime Detection (Chicago: Institute of Applied Science, 1971), lesson 1, 22-23.

6. Block, *Fingerprinting,* 4.

7. "A Study of Finger Prints," lesson 1, 23-24; Block, *Fingerprinting,* 4.

8. "A Study of Finger Prints," lesson 1, 24-25; Block, *Fingerprinting,* 4-7; Phil McArdle and Karen McArdle, *Fatal Fascination: Where Fact Meets Fiction in Police Work* (Boston: Houghton Mifflin Co., 1988), 126-30.

9. "A Study of Finger Prints," lesson 1, 23.

10. Ibid., 26-27; Brian Marriner, *On Death's Bloody Trail* (New York: St. Martin's Press, 1991), 170-71.

11. "A Study of Finger Prints," lesson 1, 25.

12. McArdle and McArdle, *Fatal Fascination,* 119-22; Stuart Kind, "Criminal Identification," in *Science against Crime,* ed. Yvonne Deutch (New York: Exeter Books, 1982), 21-23; Michael Kurland, *A Gallery of Rogues: Portraits in True Crime* (New York: Prentice Hall, 1994), 30-31.

13. Joe Nickell, "The Two 'Will Wests': A New Verdict," *Journal of Police Science and Administration,* vol. 8, no. 4 (Dec. 1980): 406-13.

14. David Fisher, *Hard Evidence* (New York: Dell, 1995), 157.

15. Wyatt Blassingame, *Science Catches the Criminal* (New York: Dodd, Mead, 1975), 12.

16. Nickell, "The Two 'Will Wests,'" 406-13.

17. Robinson, *Science Catches the Criminal*, 50.

18. "A Study of Finger Prints," lesson 1, 29; Fisher, *Hard Evidence*, 157.

19. Fisher, *Hard Evidence*, 175. See also Saferstein, *Criminalistics*, illus. 149.

20. Francis Galton, quoted in "A Study of Finger Prints," lesson 1, 31.

21. Marriner, *On Death's Bloody Trail*, 172-73.

22. McArdle and McArdle, *Fatal Fascination*, 122-23. Cf. Colin Wilson, *Clues! A History of Forensic Detection* (New York: Warner Books, 1989), 116-17.

23. Marriner, *On Death's Bloody Trail*, 175.

24. Federal Bureau of Investigation, *Classification of Fingerprints* (Washington: U.S. Government Printing Office, 1947), 13-15; "A Study of Finger Prints," lesson 3, 1-11.

25. "A Study of Finger Prints," lesson 5, 1-10; *Classification of Fingerprints*, 40-57.

26. *Classification of Fingerprints*, 27-40; "A Study of Finger Prints," lesson 3, 11-40.

27. "A Study of Finger Prints," lesson 4, 1-32; *Classification of Fingerprints*, 58-79.

28. Fisher, *Hard Evidence*, 337.

29. "A Study of Finger Prints," lesson 6, 1-26; lesson 7, 1-9.

30. Ibid., lesson 7, 10-13; lesson 5, 28-31.

31. *Classification of Fingerprints*, 96-111.

32. Ibid., 96-109; Robinson, *Science Catches the Criminal*, 47-48.

33. *Classification of Fingerprints*, 122-25.

34. Saferstein, *Criminalistics*, 423.

35. Ibid.

36. Ibid.

37. Harold Tuthill, *Individualization Principles and Procedures in Criminalistics* (Salem, Oregon: Lightning Powder, 1994), 10.

38. "A Study of Finger Prints," lesson 5, 10-14.

39. Ibid.; Charles E. O'Hara, *Fundamentals of Criminal Investigation*, 3d ed. (Springfield, Ill.: Charles C. Thomas, 1973), 672.

40. "A Study of Finger Prints," lesson 5, 15-16; see also lesson 2, 17-26.

41. Ibid., lesson 17, 6-7.

42. Charles R. Swanson Jr., Neil C. Chamelin, and Leonard Territo, *Criminal Investigation*, 4th ed. (New York: McGraw-Hill, 1988), 74.

43. Saferstein, *Criminalistics*, 416.

44. "A Study of Finger Prints," lesson 17, 6-7; Swanson, Chamelin, and Territo, *Criminal Investigation*, 74.

45. Robin Odell, "Fingerprints," in Deutch, *Science against Crime*, 119-22.

46. Eric Berg, "The Digital Future of Investigations," *Law Enforcement Technology* (Aug. 1995): 38-40.

47. "A Study of Finger Prints," lesson 17, 16.

48. Block, *Fingerprinting*, vii-viii.

49. Odell, "Fingerprints," 125-26.

50. Block, *Fingerprinting*, 37-42.

51. Ibid., 11-12.

52. Robinson, *Science Catches the Criminal*, 36-37.

53. O'Hara, *Fundamentals*, 660; Saferstein, *Criminalistics*, 425.

54. O'Hara, *Fundamentals*, 661.

55. Ibid.

56. Federal Bureau of Investigation, *The Science of Fingerprints*, rev. ed. (Washington, D.C.: Government Printing Office, 1984), 170-84.

57. Barry A.J. Fisher, *Techniques of Crime Scene Investigation*, 5th ed. (New York; Elsevier, 1992), 99.

58. *The Science of Fingerprints*, 170-71.

59. O'Hara, *Fundamentals*, 56, 662.

60. John F. Fischer and Joe Nickell, "Laser Light: Space Age Forensics," *Law Enforcement Technology* (Sept. 1984): 26-27.

61. John F. Fischer, "Forensic Light Sources and Their Use in Conjunction with Luminescent Techniques," presented at the International Forensic Symposium on Latent Prints, FBI Academy, Quantico, Virginia, 1993.

62. *The Science of Fingerprints*, 185.

63. Ibid.

64. Fischer, "Forensic Light Sources."

65. *The Science of Fingerprints*, 182-83.

66. Saferstein, *Criminalistics*, 430-31.

67. *The Science of Fingerprints*, 175, 177-79.

68. Saferstein, *Criminalistics*, 432-33.

69. *The Science of Fingerprints*, 179-82.

70. Fisher, *Techniques of Crime Scene Investigation*, 109.

71. Ivan Ross Futrell, "Hidden Evidence: Latent Prints on Human Skin," *FBI Law Enforcement Bulletin*, vol. 65, no. 4 (April 1996): 21-24.

72. Berg, *The Digital Future of Investigations*, 38.

73. Paul Walrath, "Detectives Credit Digital Imaging in Quickly Solving Cases," *Law Enforcement Technology* (Oct. 1995): 98-99.

74. Guillaume P. Vigo, Dennis M. Hueber, and Tuan Vo-Dinh, "Evaluation of Data Treatment Techniques for Improved Analysis of Fingerprint Images," *Journal of Forensic Sciences*, JFSCA, vol. 40, no. 5 (Sept. 1995): 826-837.

75. Futrell, "Hidden Evidence," 24.

76. Quoted in Michael Reynolds, *Dead Ends* (New York: Warner Books, 1992), 234. Except as otherwise noted, information in our account is based on this book.

77. Ibid., 5.

78. Ibid., 6.

79. Sam Roen, Orlando, Florida, typescript article, April 27, 1992, 1.

80. Reynolds, *Dead Ends*, 7-11.

81. Ibid., 13; *The Science of Fingerprints*, 129-36.

82. *The Science of Fingerprints*, 136-57.

83. Reynolds, *Dead Ends*, 13-14.

84. Ibid., 14.
85. Ibid., 47-104; Roen, typescript, 1-3; Fido, *The Chronicle of Crime*, 308.
86. Reynolds, *Dead Ends*, 83-112.
87. Ibid., 112.
88. Ibid., 118.
89. Deborah Fischer, personal communication with authors, July 25, 1996.
90. Ibid.; Reynolds, *Dead Ends*, 119.
91. Reynolds, *Dead Ends,* 119.
92. Ibid., 119-22.
93. Ibid., 122.
94. Ibid., 162-271; Roen, typescript, 6-15.
95. Reynolds, *Dead Ends*, 238.
96. Ibid.
97. Fido, *The Chronicle of Crime*, 308.
98. Roen, typescript, 16.

RECOMMENDED READING

Berg, Erik. "The Digital Future of Investigations." *Law Enforcement Technology*, (Aug. 1995). A nontechnical discussion of how the process of *digitizing* works and how it is applied to such forensic sciences as fingerprinting, especially the enhancement of latent prints.

Block, Eugene. *Fingerprinting: Magic Weapon against Crime.* New York: David McKay Company, 1969. A popular but now somewhat dated look at most aspects of fingerprinting, presented largely as a series of highly readable case studies with chapter titles such as "The Question Mark Burglar," "The Mystery of the Mummy," and "Telltale Letters"; also includes a history of the science, classification basics, etc.

Federal Bureau of Investigation. *The Science of Fingerprints*, rev. ed. Washington, D.C,: Government Printing Office, 1984. A comprehensive study of the science of fingerprinting that explains the basics of pattern recognition, classification, taking fingerprints, developing latent prints, etc.; tells how to establish a local fingerprint identification bureau.

Futrell, Ivan Ross. "Hidden Evidence: Latent Prints on Human Skin." *FBI Law Enforcement Bulletin*, vol. 65, no. 4 (April 1996). Discusses methods of developing latent fingerprints on the skin of crime victims, especially the most successful method that uses cyanoacrylate fuming in conjunction with magnetic fingerprint powder.

Nickell, Joe. "The Two 'Will Wests': A New Verdict." *Journal of Police Science and Administration*, vol. 8, no. 4 (Dec. 1980). New research proving that this historic case of lookalikes—which helped promote finger-printing—was not what it seemed. Far from being "unrelated," both Will West and Bill West were actually monozygotic (identical) twins. Evidence includes similar, inheritable traits in the men's fingerprints.

Vigo, Guillaume P., Dennis M. Hueber, and Tuan Vo-Dinh. "Evaluation of Data Treatment Techniques for Improved Analysis of Fingerprint Images." *Journal of Forensic Sciences,* JFSCA, vol. 40, no. 5 (Sept. 1995). A highly technical evaluation of four methods of data treatment of fingerprint images, including those to remove electronic "noise" and those that also enhance the ridge detail; illustrated.

6 IMPRESSION ANALYSIS

Just as the markings left by firearms on bullets and shell casings are distinctive and the impressions of the bare hands and soles of the feet are unique, so may other mechanical markings and other impressions be *individualized* for forensic purposes. The familiar principles apply. Remember that no two things are precisely identical and that forensic comparisons involve matching features of the questioned or unidentified specimen with known standards. In this chapter we look at *fabric prints, shoe and tire impressions,* and *tool marks.*

FABRIC PRINTS

Fabric prints are sometimes encountered by the fingerprint expert in the form of glove impressions. With use, gloves become contaminated with substances such as dirt, grease, oils, and even sweat that enable them to leave their imprints, either latent or visible. They also may leave plastic impressions in materials like putty. Such impressions are developed, photographed, and collected just as if they were fingerprints because they may be their equivalents. Cloth gloves may have snags, tears, or holes that make them distinctive, along with imperfections in the weave pattern or other identifiable characteristics. Leather gloves may have wrinkle or crease formations where they do not fit properly or they may exhibit surface cracks, tears, or other imperfections or characteristics by which they may be individualized.[1] In an English case, some unusual prints were found at the scene of a theft, and investigation soon led to a suspect who had in his possession a pair of pigskin gloves. The fingertips of the gloves were compared with the latent impressions recovered from the crime scene and showed sufficient similarities for a match.[2]

Developing glove prints requires special care because they are typically less strong than fingerprints and may be destroyed by use of too much powder. Therefore, they should first be searched for with the aid of a flashlight, laser, or other light source and then developed with sparing use of powder. Rather than lifting the print, the object on which it is found should, if possible, be taken to the lab for direct comparison with a suspect's gloves. According to one authoritative text:

> Comparison prints from the gloves of a suspect are best made on glass, which is generally the most convenient even when the original prints are on furniture. In certain special cases, however, it may be necessary to form a print on the same kind of material as that which was the scene of the crime. Where possible, such material should always be enclosed when a print and gloves from a suspect are sent for examination. Comparison prints should be made in a manner similar to the original ones. Thus if it is possible, in view of the placing of the original, to decide, for example, how the hand of the suspect gripped when making the grip, this information should be communicated to the expert so that the same grip can be used for the comparison print. Consideration must also be given to the degree of pressure that may have been used in forming the original print, and statements forming a guide for judging the pressure should also be submitted. It is of great importance that neither too great nor too small a pressure should be used in making the comparison prints since their appearance is greatly affected by pressure.[3]

Clear prints may be difficult to make, but results may be improved by breathing slightly on the surface of the glove's finger. Caution should be used in applying powder, leather polish, or the like in order to avoid destroying subtle details.[4]

Apart from glove prints, other types of fabric impressions may be found at the scene of a crime under certain circumstances. One of these is the hit-and-run accident in which fabric impressions from the victim's clothes (along with actual fibers, blood, hair, etc.) may be left in the dust on the assault vehicle's bumper, grille, fender, or other area. A hydraulic lift should be used so that the undercarriage may also be examined.[5]

Whether latent, visible, or plastic, fabric prints should be treated in the manner recommended for glove prints. In photographing them, the camera should be at right angles to the surface to avoid distortions. Photographing should be done both with and without a scale, as discussed in chapter 2 in the section on documentation. If such prints are to be lifted, whether found in light deposits of dust or dirt or after being dusted with fingerprint powder, that may be accomplished with sheets of lifting material like that used for fingerprints. There is a new procedure that involves a portable electrostatic device operating on a principle similar to that of using the static charge on a comb to lift bits of paper. A special plastic lifting film is placed over the imprint and charged from a power supply, causing the dust or powder to be attracted to the lifting film. (The attachment to the film is not permanent, so the imprint is photographed for preservation.) This technique is especially useful in recovering faint prints from colored surfaces.[6]

When used as evidence, fabric impressions may show sufficient *class* characteristics to demonstrate that a particular garment *could* have made the print, or they may exhibit such individual characteristics arising from wear or damage that the examiner can conclude the garment *did* leave the impression.[7] Even in the former instance, there is evidential value that—taken with other physical or circumstantial evidence—may prove helpful in winning a case.

SHOE AND TIRE IMPRESSIONS

Shoe prints and tire prints are left in much the same way as are fabric and other types of impressions. The shoe or tire becomes contaminated with some substance such as blood, oil, or dust, and a latent or visible print is left; alternatively, the print may be a plastic or three-dimensional one, pressed into mud, for example.

Latent shoe prints are searched for and developed in much the same way as glove and other fabric prints. Whenever the impression is made by dust or other powdered material, the procedure is first to photograph it—oblique lighting is usually helpful—and then to lift it in the manner suggested for fabric prints; that is, by using sheets of lifting material or by the electrostatic technique. Comparison footprints may be made by coating the soles with water-based ink applied from a large ink pad; the shoes are then stepped onto an acetate sheet or a sheet of tracing paper.[8] When the contaminating substance is blood—as in the case of a completely invisible shoe print left at the scene of a double homicide in Nome, Alaska—it may be developed by spraying with the chemical reagent luminol. The result is a faintly glowing print that must be observed in darkness. In the Nome case, the developed and photographed print was successfully matched with a test impression made from the shoe of the suspect.[9]

When either a shoe or tire impression has been left in some moldable substance such as clay or snow, it is first photographed from directly above to avoid distortion, and then it is cast. Traditionally, plaster of paris has been used to cast shoe and tire impressions, but in recent times dental stone has been found superior. In either case, because the weight of the soupy material may damage or distort the impression, the surface is first carefully sprayed with shellac or clear acrylic lacquer. With imprints in snow, a special Snow Print Wax is used instead. Two coats of the proper preservative are applied, with sufficient drying time allowed between applications. Next, a retaining frame with a depth of two inches is placed around the impression, and the casting material is carefully poured in. (In the case of plaster of paris, the cast is reinforced by placing wire mesh over the print after the pouring reaches a depth of about half an inch, followed by the deposit of the remaining one and one-half inches of poured plaster.) Before the plaster or dental stone has hardened, identifying information is scratched into the surface.[10]

If the impression is clear, examination of the recovered shoe print or tire print may permit determination of the manufacturer of the shoe or tire from the class characteristics.[11] The FBI maintains a vast file of both shoe (sole and heel) patterns and tire-tread designs for such identifications.[12] For example, during the murder trial of O.J. Simpson, the prosecution introduced evidence identifying bloody shoe prints at the crime scene as having been made by size-twelve Bruno Magli shoes—designer footwear that is both expensive and rare. After Simpson's astonishing acquittal, a photograph surfaced showing Simpson wearing just such Bruno Magli shoes.[13]

In addition,

> Tire tread impressions are almost as useful as shoe prints. There are thousands of tread designs, and every tire wears differently or picks up debris from the road, making it unique and therefore identifiable. Tire treads are made of what are called elements, which are small sections of the tire with a specific design that is repeated over the entire tire. Some tires have as few as three different elements, while better tires have as many as nine. These elements are designed to improve traction and cut down on the noise the rubber makes as it is squeezed against the road surface. Tread design—the size and location of the elements—is what makes a tire identifiable. Sci-crime detectives can not only identify

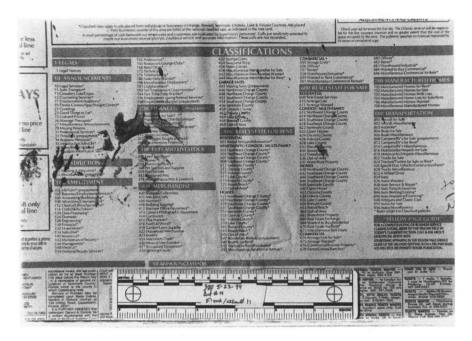

FIGURES 6.1, 6.2, AND 6.3. A partial bloody shoe impression on a newspaper recovered from a homicide scene (fig. 6.1) was examined using a shortwave ultraviolet light, which revealed the remaining portion of the shoe impression (fig. 6.2). The sole of the shoe recovered from the suspect (fig. 6.3) was successfully matched with the ultraviolet light-enhanced photograph.

a specific tire from tread impression, they can often find the portion of the tire that left the impression.[14]

In making comparisons between shoe or tire prints found at a crime scene and test imprints, it should be kept in mind that dimensions may not exactly match due to stretching of the imprint from motion, distortion due to pressure, shrinkage caused by drying of earth or clay, or other factors. What is important is that there be discovered points of comparison—the various nicks, scratches, and other effects of wear—sufficient to effect a match. For presentation of the evidence at trial, photographs of the questioned and known prints are placed side by side, with the characteristic points marked and numbered rather like the minutiae of fingerprints.[15]

The following case will demonstrate the value of impression evidence in solving crimes. It was brilliantly handled by the FBI's top footwear expert, William Bodziak:

FIGURE 6.2.

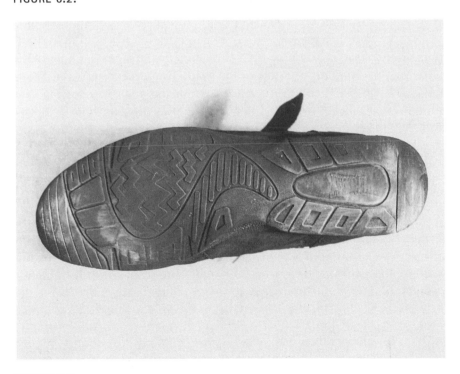

FIGURE 6.3.

When a six-year-old Florida girl was brutally raped and murdered, Bodziak was able to find evidence in the bruises the killer left on her skin. "There was a contusion, a bruise, made by the heel of the killer's shoe when he kicked her," he said. "Normally contusions will give you some detail, but they rarely show the small details, cuts and scratches in the shoe, that enable us to match it to one shoe to the exclusion of all others. But this was unusual. I could determine the size and the design of the shoe, and I could even see evidence of wear. I was able to determine that the contusions had been made by a deck shoe whose heel was worn down, and I could see precisely where the stitching had gone into the heel. Stitching is done randomly; no two shoes are alike. Using the stitching holes, I was able to match the impression on that little girl's skin to the suspect's shoe. It was the footwear impression that really got us the conviction."[16]

TOOL MARKS

Tool-mark examinations may have begun in Europe around the turn of the century. They became common in the United States in the 1930s following the successful use of the comparison microscope for matching striations on bullets and firing-pin impression marks on shell casings. Indeed, the methodology used for firearm examination and that for tool-mark comparison is fundamentally the same. For that reason, tool-mark matching has tended to be taken for granted. Nevertheless, the 1950s brought a new focus on this specialty, including a study involving one hundred sequentially manufactured chisels. Over five thousand comparisons were made, demonstrating that a tool mark produced by any given chisel is distinct from marks produced by all other chisels.[17]

The basis of tool-mark comparisons is the fact that implements used for cutting metal or for prying apart objects develop nicks on their edges with use. These nicks produce a pattern of striations during the cutting or prying process that is distinctive for a given tool. By using the suspected tool to make a test mark on a plate of relatively soft material such as lead, the examiner may then compare that known mark with the questioned one from the crime scene, as is done in other forensic comparisons. The comparison microscope is used in a technique that is essentially the same as for bullet comparisons. Marks from chisels, crowbars, screwdrivers (used for prying), and similar implements are matched in this manner. The science involved is called "comparative micrography."[18] Cutting tools such as bolt or wire cutters present a special problem. Since the cutting blade is long in comparison to the cut region of the item involved (padlock shackle, electric wire, etc.), the examiner will want to avoid having to make numerous test cuts in order to find the cutting region. One way to accomplish this is to use chemical spot tests to identify traces of the cut material on the tool's blade. In the case of an attempted burglary, for example, a chemical reagent that turns red-violet in the presence of zinc can be employed in identifying the region of a bolt cutter's blades used to cut the zinc-coated steel shackle of a padlock.[19]

To make a comparison of striations, the tool-mark area—for example, the severed end of the padlock shackle or link of a chain, or the scrape mark of a crowbar—is placed under one barrel of the comparison microscope, and the cut or scraped area of the lead test plate is placed under the other. If a number of the striations of each match, the expert may be able to conclude that the tool that marked the test plate is the same tool that left its mark at the crime scene. There may be some differences due to bending of the metal during cutting or by application of uneven pressure or other factors. For this reason, since lay persons viewing the photomicrographs may be confused, some experts avoid using the illustrative exhibits and instead simply explain their test procedures and conclusions.[20]

Just as some tools leave striations that are comparable to those on a bullet, others may simply leave an indentation or impression that is similar to a firing pin mark on a shell case. A hammer blow would represent this type of mark, as would the indentations caused by a pair of pliers. Again, the comparison microscope is used to compare the outer shape of the impressions as well as any surface scratches, nicks, pits, or other defects. This is another procedure of comparative micrography. There may even be a combination of marks, both striations and indentations. Comparative micrography techniques also may be employed in matching a piece of broken metal or other material (found, say, at the scene of an auto accident) and the piece it was broken from (such as a remnant on a suspected vehicle).[21]

Even when a suspect's tool has not yet been recovered, comparison micrography may reveal that tool impressions at a series of burglaries were caused by the same tool, thereby linking the crimes. Once tool marks are discovered, they must be protected until they can be photographed and their locations indicated on the crime-scene sketch. (Two photos are required: one showing the mark in its context, the other a close-up photograph to reveal the minute details.) When warranted and practical, the structure bearing the tool mark is removed. Otherwise, a cast or mold may be made with plasticine or modeling clay.[22] The authors of *Criminal Investigation* caution:

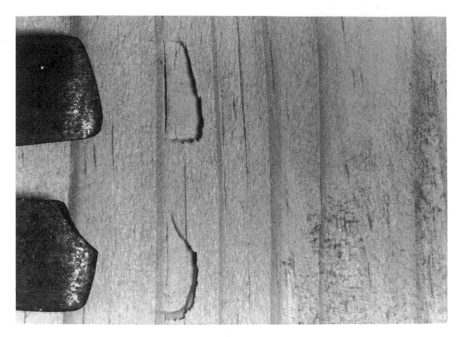

FIGURE 6.4. Comparison of a set of toolmarks recovered from a crime scene with the suspected tool (a claw hammer) that made it. The nicks and especially the broken corner of the claw help to individualize the tool to the mark.

In no event should the investigator place a tool against a tool mark for size evaluation, as it may lead to accidental destruction of evidence. When submitting a tool to the crime laboratory for examination, the actual tool should be submitted; the making of test impressions or cuts are functions of qualified technicians in the laboratory. The importance of this last point is illustrated by the fact that under test conditions in the laboratory it was found that when there was more than 15 degrees difference between the vertical angle at which a screwdriver was actually used and the comparison mark made in the laboratory, an improper finding of no identity from the same tool could result.[23]

Therefore, a series of test marks should be made, applying the suspected implement at varying angles and pressures to the lead plate. This increases the likelihood of duplicating the details of the original mark.[24]

The following case study explains how comparative micrography of tool marks was employed to trace the ladder used in the Lindbergh kidnapping. The case gave prominence to the science of tool-mark com-

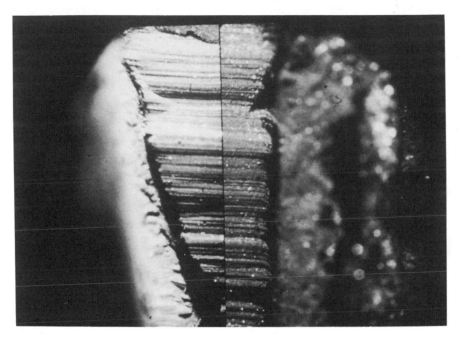

FIGURE 6.5. Test marks taken from tools found in possession of burglary suspects can be compared and often matched to toolmark impressions recovered from crime scenes. Such a match is shown in this photomicrograph from a comparison microscope. (Courtesy of Thomas Fadul, Metro-Dade Police Department, Miami, Florida.)

parisons in much the same way that the St. Valentine's Day Massacre did for firearms identification.[25]

CASE STUDY: THE LINDBERGH KIDNAPPING

It was the night of March 1, 1932. At the home of famed aviator Col. Charles A. Lindbergh outside of Hopewell, New Jersey, all seemed well. His wife, poet Anne Morrow Lindbergh, the daughter of a wealthy ambassador, had put to bed the couple's twenty-month-old son, Charles Junior, at about eight o'clock. The little boy's nurse had stayed with him for the few minutes it took him to fall asleep.[26]

Colonel Lindbergh had been the first person to fly an airplane solo across the Atlantic. In his *Spirit of St. Louis*, he had taken off from Mineola, New York, on May 20, 1927, and had landed in Paris the following day. He became a national hero, dubbed "the Lone Eagle." His prominence also made him a target. At nine o'clock, Mrs. Lindbergh had looked in on her "little It"—as she and her husband called their baby—and found him sleeping peacefully in his crib. Just fifty minutes later, the nurse discovered him missing. As Colonel Lindbergh instructed the butler to telephone the police, he grabbed a Springfield rifle and searched the immediate grounds of his estate. He found nothing more than a wooden ladder lying beneath the nursery window. On the upstairs sill was an envelope that Lindbergh left unopened until the police arrived. And arrive they did: a "foaming and senseless cataract of gorgeously uniformed state troopers that descended on the Lindbergh home in motorcycles, roared up and down the road trampling every available clue into the March mud, systematically covering with impenetrable layers of stupidity every fingerprint, foot print, dust trace on the estate."[27]

The police found few clues. The ladder was homemade, built in three sections for ease of carrying. An upper rung was broken, apparently having given way under the kidnapper's weight. After midnight, a fingerprint expert arrived and dusted the envelope for latent prints. There were none. In tortured handwriting was written:

> Dear Sir!
> Have 50000$ redy 25000$ in
> 20$ bills 15000$ in 10$ bills and
> 10000$ in 5$ bills. After 2-4 days
> we will inform you were to deliver
> the mony.
> We warn you for making
> anyding public or for notify the Police.
> the child is in gute care.
> Indication for all letters are
> signature
>   and three holds.[28]

Unfortunately, the police were everywhere, and so was news of the fact.

A profile of the kidnapper was emerging, however. The ransom note indicated a man of poor education, almost certainly of German origin. If there was only one kidnapper, he appeared to have tools and some knowledge of carpentry. Soon several lines of inquiry pointed to the Bronx.

Colonel Lindbergh and his friends and colleagues, including Col. H. Norman Schwarzkopf, superintendent of the New Jersey Police, thought the kidnapping might have been the work of mobsters, and he enlisted two prospective intermediaries who had underworld connections. At the same time, a self-appointed well-wisher named John F. Condon, a seventy-two-year-old public school principal, sent a letter to the *Bronx Home News* offering a thousand-dollar reward for the return of the child. His reward was a far cry from the reward offered by the Lindberghs (fifty thousand dollars) or the State of New Jersey (twenty-five thousand dollars), but Condon's published offer received an unexpected reply. It bore the same interlocking rings and punched holes as the ransom note and stated that Condon would be acceptable as an intermediary. He rushed to Colonel Lindbergh, who gave his go-ahead. Condon placed an ad as requested by the kidnapper, signing himself "Jafsie" (from the pronunciation of his initials).

Jafsie soon received a phone call arranging for a meeting at night in the Bronx's Woodlawn Cemetery. At the gates, a man disguised by a handkerchief held to his face asked Condon if he had the cash. Condon replied that the money was not yet ready. Something spooked the man, who began to run away, but Condon chased after and caught up with him, assuring him that he had nothing to fear. Identifying himself as "John," the man asked abruptly: "Would I burn if the baby is dead?" "*Is* the baby dead?" Condon asked. The man replied that the child was alive and well, being kept on a boat. He promised to send the baby's sleeping suit as a token of good faith. When it arrived, Lindbergh identified it as the one his son wore when kidnapped.

On April 2, Condon prepared to hand over the ransom money, which was packed in a box that was specially made for later identification. That night a man hailed a cab in the Bronx and gave the driver a note to Jafsie. Lindbergh was with Condon when he received the instructions, and the pair proceeded to a flower shop where, as instructed, they found a note under a stone. It set the rendezvous at St. Raymond's Cemetery. There, with Colonel Lindbergh within earshot, Jafsie exchanged the box for a note, and the man disappeared among the tombstones. The note claimed that the baby was being held off Martha's Vineyard on a boat named *Nelly*. Lindbergh flew there and searched fruitlessly for two days.

On May 12 the decomposing body of a child was found in woods near the Lindbergh home. It was identified by Lindbergh as his son's remains. The child had reportedly died from a head injury—possibly suffered in the kidnapper's fall from the ladder. (In later years, forensic review by famed medical examiner Dr. Michael M. Baden found that there had been no skull fracture or brain injury. Rather, according to Baden, "The baby was probably smothered at the time of the kidnapping to keep him from crying out and alerting the family and the nurse who were all in nearby rooms.")[29]

Before finding the body, the police had largely been kept at bay, out of fear for the child's safety. With that no longer an issue, the investigation proceeded unhindered. Although Lindbergh had refused to allow the money to be marked, the police had secretly recorded the serial numbers, and now banks were alerted to look for the bills. From time to time, one turned up in New York or Chicago. In 1933 the United States abandoned the gold standard and required people to turn in their old bills before May 1. Not long before the deadline, a man signing himself as "J.J. Faulkner" redeemed $2,990 of the ransom money at the Federal Reserve Bank in Manhattan. Unfortunately, the bank staff was so busy at the time that no one could recall what the man looked like.

Lindbergh ransom bills began turning up in the Bronx, but they were invariably identified only after it was too late to trace them. In November 1933, a ticket seller at a Loews theater in Greenwich Village became suspicious when a man tendering a five-dollar bill acted in a "furtive" manner. She thought the bill might be counterfeit but soon discovered that it was part of the ransom booty. Unfortunately, the man was gone by the time police arrived. It would be ten months before they got the break they needed.

Again, "furtive" behavior was noted, this time by a Bronx service station attendant who took down the license plate number of the dark-blue 1930 Dodge sedan the man was driving. The attendant wrote the number directly on the back of the customer's ten-dollar gold certificate. Three days passed before a bank teller recognized it as one of the ransom bills. On the back was penciled "4U-13-41, N.Y.;" the number of a license plate registered to one Richard Hauptmann, 1279 East 222nd Street, the Bronx. Quickly, but cautiously, police infiltrated the neighborhood and worked through the night to gather information on Hauptmann and to decide whether to maintain a surveillance (hoping he would lead them to accomplices) or to arrest him. Finally a decision was made.

At eight-fifteen on the morning of Wednesday, September 19, 1934, Hauptmann left his home and drove along Park Avenue until police—growing fearful that he might escape—finally acted. One of their three black, unmarked Fords pinned in Hauptmann's Dodge sedan while another came alongside and the third drove in behind to block the rear. Troopers with guns drawn pulled Hauptmann from the car, handcuffed him, and led him to the sidewalk, where he was searched. He was unarmed, but among the several bills in his wallet was a twenty-dollar gold certificate that had come from the ransom money.

Not only was he a German and a carpenter who fit the description of the man "Jafsie" had rendezvoused with, not only had he been passing Lindbergh ransom money, but there was further evidence awaiting the police at Hauptmann's residence, a second-floor apartment. In a small frame garage at the edge of the property was a workshop area containing a carpentry bench. In two hidden shelves in the garage were packages of money totaling $13,760—all of it from the ransom payment. Another $840 was eventually found, along with a tiny handgun. Police estimated Hauptmann had placed twenty-five thousand dollars in the stock market and had spent sufficient additional money since he quit his job—on the day of the ransom payment!—to account for virtually all of the ransom money.

Hauptmann was taken to the NYPD's Second Precinct station and fingerprinted. Under interrogation he told several lies, one when he denied having more gold certificates at home and another when he denied having a criminal record in Europe. Soon, witnesses brought to the station identified Hauptmann, one of them the cab driver whom Hauptmann had given a dollar to deliver a ransom note to "Jafsie." Condon himself also identified Hauptmann as "John," although he was initially unsure. He did pick him out of a lineup of fourteen men as the most likely suspect, saying Hauptmann resembled the man to whom he had given the ransom money. (Condon had only glimpsed "John's" covered face in the dark and had briefly exchanged a few words with him.) The theater cashier and several others identified Hauptmann as the passer of Lindbergh ransom bills.

Some evidence against Hauptmann was disputed. Condon's telephone number was written on a wooden beam inside a closet at Hauptmann's home, but the penciled notation was subsequently credited to a reporter for the *New York Daily News*. Also, a poor neighbor of the Lindberghs, who had first told police he had seen nothing, had since picked Hauptmann out of a lineup as a man seen in the area of the estate in the days before the kidnapping.[30]

Soon, it would be forensic science that would provide the most powerful evidence against Hauptmann. Handwriting samples obtained from the suspect were matched to the writing of the ransom notes. This evidence (which will be discussed in the following chapter) is so solid that it has been confirmed over and over by handwriting experts and is still used in training examiners of questioned documents.

Other impressive forensic evidence came from the marks of the tools used to make the ladder that was left at the kidnapping scene. The ladder was positively traced to Bruno Richard Hauptmann's workshop. This forensic work was accomplished by "wood expert" Arthur Koehler, who was said to have "wrapped the kidnap ladder around Hauptmann's neck."[31] Koehler was a carpenter and son of a carpenter who obtained both a bachelor's and master's degree in forestry and became the federal government's top expert on wood technology and identification. He was stationed at the United States Forest Products Laboratory in Madison, Wisconsin. According to one account:

> Koehler's entrance into the Lindbergh case in May, 1932, was accompanied by no fanfare of trumpets; at that time he was a very humble private in the army of investigators assigned to that crime. Bruno Richard Hauptmann was still at large and his name was utterly unknown to Koehler or anyone else when the government expert started tracing the materials used in making the kidnap ladder. Other scientists had viewed this wooden exhibit and had pored over it for indications of its maker's identity; they had daubed it with silver nitrate in an attempt to find fingerprints; they wrapped it in blankets and shipped it all over the countryside to more or less informed "experts," who, after scrutinizing it blankly, had shipped it back in its woolly blankets, as mute and as impenetrable as ever.[32]

The kidnapper had probably not intended to leave the ladder behind. But during the abduction, with the added weight of the child, a rung had broken. Colonel Lindbergh had himself heard the crashing sound—like that of "a falling crate"—which must have alarmed the kidnapper and caused him to abandon the ladder in haste.

Koehler's initial observation was that the three-sectioned ladder was made of four kinds of wood. North Carolina pine was used for the rails or uprights, Ponderosa pine for the rungs, and birch for the dowel pins that held the sections together. An odd piece of wood—apparently cut from a strip of fir flooring—had been used for the upper left section of rail, the carpenter evidently having run out of material at that point and being forced to improvise. That the board had been used before was obvious from four nail holes in it.

Koehler next used his microscope and a special type of oblique light to discover the marks of the planing machine used to dress the side rails. Koehler discovered that most of the rails had been dressed with a plane whose top and bottom cutter heads had eight knives, one with a nick and one out of alignment, and whose side cutter heads had six blades, one of which had a characteristic defect. From the two manufacturers of planing machines, the wood detective compiled a list of 1,595 mills across the country that had the machines. By asking them to send samples of planed one-by-four pine stock, Koehler finally discovered one bearing the same distinctly individualized plane marks as those on the rails of the kidnap ladder. It came from the Dorn lumber in McCormick, South Carolina, which had sold North Carolina pine to twenty-five companies.

Eventually Koehler traced the wood that had been used in the kidnap ladder to the National Lumber and Mill Work Company of the Bronx. Unfortunately, after eighteen months of searching, Koehler learned that National Lumber had no list of persons who had bought the one-by-four pine stock. The company made only cash sales and kept no itemized sales records. So Koehler would be unable to locate the kidnapper as he had hoped but would instead have to await a suspect's arrest in order to proceed further.

As it happened, upon his arrest Bruno Richard Hauptmann admitted that he was formerly employed by National Lumber. Moreover, a cash sales slip showed Hauptmann had made a ten-dollar purchase of unitemized lumber at the yard in December 1931. Koehler searched Hauptmann's premises and, a half hour later, had found in the attic a floor joist containing four nail holes. These corresponded exactly to the four holes in the upper rail of the kidnap ladder. Koehler also matched the grain and annual rings in the attic board and the ladder rail. According to Robinson in his *Science Catches the Criminal*:

> The final straw in the weight of evidence was the testimony of Hauptmann's own plane. With this tool the kidnapper had smoothed down the material in the ladder, and in so doing had left unmistakable marks of the plane's identity on every cleat and rail. When Koehler laid hands on Hauptmann's plane which he found on a dusty shelf in the garage, the wood expert trembled with the fierce delight that comes to a man when he is on the verge of solving a difficult problem. He could scarcely wait to shove its dull edge—unsharpened and neglected ever since Hauptmann had secured the big ransom payment—across a piece of lumber. Koehler knew exactly the marks that he wanted the plane to make; he had examined those marks on the kidnap ladder a thousand times, always noting the scuffed ridges left by the abused tool. To make these distinctive marking visible to others he resorted to a trick well known to every school child. Fixing a sheet of paper over the plane marks on the ladder-rail he rubbed a lead pencil back and forth across the surface of the paper until a black and white diagram of the plane's imperfections was transferred to the white sheet. Then taking a fresh piece of lumber he pushed Hauptmann's plane across its surface. Now for a second time he got out a sheet of paper, placed it over the newly planed

board and rubbed it as before with his lead pencil. Comparing the two pieces of paper he could easily demonstrate that they contained the same markings, the same ridges and longitudinal striations! This he demonstrated to the Flemington jury and the case against the ladder-maker was complete. The ragged track of that dull-edged plane probably did more to convict Hauptmann than all the rest of the evidence combined.[33]

Hauptmann's trial for kidnapping and murder began on January 2, 1935, in Flemington, New Jersey. In addition to the testimony of those already mentioned, Colonel Lindbergh testified that Hauptmann was the person whose voice he had overheard when he accompanied "Jafsie" to make the ransom payment in the cemetery. On February 13, Hauptmann was convicted and sentenced to die in the electric chair. His petition for clemency was rejected, and the execution was set for January 17. Various postponements extended it to April 3. Hauptmann protested his innocence to the end, even though the governor promised him he could save his life by telling the truth. At 8:47 P.M. on the appointed date, Bruno Richard Hauptmann was pronounced dead.

Even today, the case remains controversial. Small wonder, with authors misinforming their readers as Michael Kurland does in his *A Gallery of Rogues* when he states, "The only thing we can be sure Hauptmann was guilty of was possession of some of the ransom money."[34] In fact, those who can understand the forensic evidence are able to see the depth of Hauptmann's guilt and to know that justice was done.

NOTES

1. Barry A.J. Fisher, *Techniques of Crime Scene Investigation*, 5th ed. (New York; Elsevier, 1992), 118-20.

2. Stuart Kind and Michael Overman, *Science against Crime* (Garden City, N.Y.: Doubleday, 1972), 34.

3. Fisher, *Techniques*, 120-21.

4. Ibid., 121.

5. Ibid., 390-91.

6. Richard Saferstein, *Criminalistics: An Introduction to Forensic Science*, 5th ed. (Englewood Cliffs, N.J.: Prentice Hall, 1995), 464-65.

7. Ibid., 465-66.

8. Fisher, *Techniques*, 251.

9. Ibid., 224-25, 254.

10. Charles R. Swanson Jr., Neil C. Chamelin, and Leonard Territo, *Criminal*

Investigation, 4th ed. (New York: McGraw-Hill, 1988), 60-63; Fisher, *Techniques,* 247-49.

11. Charles E. O'Hara, *Fundamentals of Criminal Investigation,* 3d ed. (Springfield, Ill.: Charles C. Thomas, 1973), 573.

12. C.B. Colby, *F.B.I.: The 'G-Men's' Weapons and Tactics for Combating Crime* (New York: Coward-McCann, 1954), 22, 23.

13. Clifford Linedecker, *O.J. A to Z: The Complete Handbook of the Trial of the Century* (New York: St. Martin's Griffin, 1995), 39; Michael Fleeman, "Shoe Photo of Simpson Scrutinized," *Buffalo News,* May 26, 1996.

14. David Fisher, *Hard Evidence* (New York: Dell, 1995), 282.

15. Fisher, *Techniques,* 253; Saferstein, *Criminalistics,* 462.

16. Fisher, *Hard Evidence,* 278.

17. Eliot Springer, "Toolmark Examinations—A Review of its Development in the Literature," *Journal of Forensic Sciences,* JFSCA, vol. 40, no. 6 (Nov. 1995): 964-68.

18. Fred E. Inbau, Andre A. Moenssens, and Louis R. Vitullo, *Scientific Police Investigation* (New York: Chilton, 1972), 87-88.

19. Yehuda Novoselsky, Baruch Glattstein, and Nikolai Volkov, "Microchemical Spot Tests in Toolmark Examination," *Journal of Forensic Sciences,* JFSCA, vol. 40, no. 5 (Sept. 1995): 865-66.

20. Inbau, Moenssens, and Vitullo, *Scientific Police Investigation,* 88-91.

21. Ibid., 92-94; O'Hara, *Fundamentals,* 708-9; Swanson, Chamelin, and Territo, *Criminal Investigation,* 105.

22. O'Hara, *Fundamentals,* 710-11.

23. Swanson, Chamelin, and Territo, *Criminal Investigation,* 107-8.

24. Saferstein, *Criminalistics,* 460-61.

25. Springer, "Toolmark Examinations," 965.

26. Except as otherwise noted, information for this case was taken from the following sources: Michael Kurland, *A Gallery of Rogues: Portraits in True Crime* (New York: Prentice Hall, 1994), 223-34; Brian Marriner, *On Death's Bloody Trail* (New York: St. Martin's Press, 1991), 225-30; and Noel Behn, *Lindbergh: The Crime* (New York: The Atlantic Monthly Press, 1994).

27. Henry Morton Robinson, *Science Versus Crime,* quoted in Kurland, *A Gallery of Rogues,* 224.

28. Transcribed from a photograph of the letter in Behn, *Lindbergh* (6th illustration following p. 208). Behn's own transcription (23-24) has several errors but is far better than Kurland's (*A Gallery of Rogues,* 224). Marriner (*On Death's Bloody Trail*) states incorrectly that the letter was "written in block capitals."

29. Michael M. Baden, *Unnatural Death: Confessions of a Medical Examiner* (New York: Random House, 1989), 6.

30. For a discussion, see Behn, *Lindbergh,* 225-29.

31. The account of Koehler's work is mostly based on Henry Morton Robinson, *Science Catches the Criminal* (New York: Blue Ribbon, 1935), 185-94.

32. Ibid., 187.

33. Ibid., 193-94.
34. Kurland, *A Gallery of Rogues*, 233.

RECOMMENDED READING

Behn, Noel. *Lindbergh: The Crime*. New York: Atlantic Monthly Press, 1994. A typical conspiracy-theory interpretation of the Lindbergh case, suggesting that the kidnapping was a cover for murder by a "member of the Lindbergh's immediate family circle—which means an innocent man was executed." Recommended only as an example of the conspiracy genre.

Fischer, Barry A.J. "Impression Evidence." In *Techniques of Crime Scene Investigation*, 5th ed., New York: Elsevier, 1992. A good forensic overview of tool and other impression marks.

Robinson, Henry Morton. "Clues in Wood." In *Science Catches the Criminal*, New York: Blue Ribbon Books, 1935. Covers various types of wood evidence including toolmarks—notably those on the ladder in the Lindbergh kidnapping.

Springer, Eliot. "Toolmark Examinations—A Review of its Development in the Literature." *Journal of Forensic Sciences*, vol. 40, no. 6 (Nov. 1995). Reviews the history and development of toolmark examinations. Extensive references.

7

QUESTIONED DOCUMENTS

Forged checks, "poison pen" letters, ransom notes, disputed legal documents, altered ledgers, counterfeit identification papers—these and similar fakes are the targets of the forensic questioned-document examiner.

Most of the examiner's methods fall under the heading of *handwriting* and *typewriting comparison* or *forgery detection* and employ various *analytical techniques*. This chapter's case study involves the "Mormon Forgery Murders."

HANDWRITING COMPARISON

The ancient Jews apparently took the first step toward the development of the science of handwriting comparison by recognizing the individuality that is inherent in handwriting. This recognition was codified in the Jewish Mishnah in the first and second centuries.[1] Unfortunately, the individuality of handwriting also has served as the basis of another type of "handwriting analysis" called *graphology*, a pseudoscience that involves the supposed divining of personality from handwriting.[2] As British expert Wilson R. Harrison notes, graphological methods have little if any experimental foundation,[3] but the work of the questioned-document examiner is established on sound principles: it relies on empirical principles, is self-correcting, and is self-policing (involving peer review and other policing mechanisms).[4]

The first time that handwriting was crucial to a widely famous case was during the Dreyfus affair of 1894-95. French army officer Alfred Dreyfus was arrested, tried, and convicted of spying on the basis of a forged letter that purportedly indicated his treason in giving important military secrets to Germany. Unfortunately, the great criminalist Alphonse Bertillon, whose testimony was instrumental in securing Dreyfus's conviction, lacked expertise in the field of handwriting comparison. Instead he tried to apply principles from his own science of anthropometry. Only after twelve years of controversy was Dreyfus declared innocent and freed from prison.[5]

The first major role that handwriting played in American history was in the "trial of the century," the Lindbergh kidnapping case, which was discussed in chapter 5. In addition to the wealth of other evidence, the ransom letters proved to have been in the "highly personalized" handwriting of Bruno Richard Hauptmann. According to Ordway Hilton in his *Scientific Examination of Questioned Documents*, "An invented form of the *x,* simplification of the *t* design ("tit for "did") . . . and the double *p* written in Hauptmann's rugged manner of execution in combination with many other habits left no question that he wrote the ransom letters." (Other unusual traits were a distinctive *not* written as "note," with an open *o* and uncrossed, tent-shaped *t;* a *to* that resembled a *w;* and a bizarre form of *the* that looked like "hle.")[6] The writing evidence against Hauptmann was assembled by Charles Appel, the founder of the famous FBI crime lab, which opened in 1932. According to one source, "Although Hauptmann's conviction remains one of the most controversial decisions in American judicial history, Appel's handwriting comparison is considered so strong that it is still used in the training of document examiners."[7]

In the thirty-five years between the Dreyfus and Lindbergh cases, experts in handwriting identification increasingly testified in American courts, particularly in the eastern United States. In 1894 came the first significant modern text that attempted a thoroughly scientific approach to questioned documents, including chemical tests, for detecting alterations—E.E. Hagan's *Disputed Handwriting.* It was followed by Albert S. Osborn's monumental *Questioned Documents* in 1910.[8] The status of the expert has continued to rise to the present time, when document examiners are thoroughly trained, carefully tested, and professionally certified by forensic bodies such as the American Society of Questioned Document Examiners.

Many similarities may exist between two or more examples of writing by different individuals because their authors learned penmanship from the same writing system. According to a forensic instruction manual:

When a child first begins to learn the art of handwriting, penman-ship copy books or blackboard illustrations of the different letters are placed before him and his first step is one of imitation only, by a pro-cess of drawing. The form of each letter at first occupies the focus of his attention. As he progresses, the matter of form recedes to the margin of attention, and finally to the subconscious mind. Then the attention is centered on the execution of the various letters—that is, they are actually written instead of drawn. Soon this manual operation likewise is relegated to the subjective mind and the process of writing becomes more or less automatic. Then and not until then, the subject matter to be written occupies the focus of attention.

This means that the particular style of penmanship learned in early childhood leaves an impress upon the mind which influences greatly the writing of later years. The mature writing is of course modified by other factors, such as education, training, personal taste, artistic ability, musculature, nerve tone, and the like; but once the form of the letters and their manual execution have been crystallized by long usage, the identifying characteristics will undergo but slight if any change as time goes on.[9]

Therefore, the writing system one learns yields *class characteristics,* and the distinctive or peculiar features that develop subsequently and that are not common to any group are *individual characteristics.* For example, a modern copybook script *a* is closed at the top and the final downstroke retraces the upstroke; those features are class characteristics. An *a* that is markedly open at the top and whose final downstroke combines with the preceding stroke to form a loop would exhibit individual characteristics.[10]

The prevalence of individual characteristics is the basis of forensic identification of handwriting. According to the great pioneer Osborn: "Only a small proportion of the vast variety of forms in writing can be accounted for by tracing them back to a parent system. Thousands of these characteristics are individual inventions and developments."[11] To see just how unique handwriting might be, the United States Postal Labo-ratory launched a project in which five hundred sets of handwriting of both fraternal and identical twins were studied. Six experienced exam-iners studied the sets of handwriting, concluding that the differences in writing characteristics between the various sets of twins were just as dis-tinctive and individualistic as would be expected between nonrelated persons.[12]

As with other forensic sciences, the basic approach with handwriting is to compare that which is unknown with what are termed known standards.

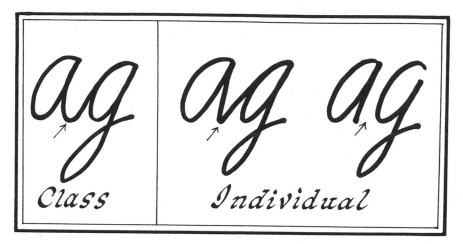

Class   Individual

FIGURE 7.1. In handwriting comparison it is essential to distinguish between *class* characteristics (such as normal curvature in connecting strokes, left) and *individual* characteristics (such as the notably angular or even absent connectors shown at right). Only the latter are used to compare one writing sample with another in the attempt to discover common authorship.

(Any known specimen used for handwriting comparison is called a "standard." The older term "exemplar" may be used to designate standard writing that is offered in evidence or obtained on request for comparison purposes.) It is important that standards be as similar as possible to the questioned writing—for example, a questioned signature with other signatures, or a ball-point scrawl or carefully written fountain pen specimen with similarly written examples. Standards should also be relatively contemporaneous to the questioned writing and contain similar words and letter combinations. They must be authentic. (In one case, endorsements on some paychecks were belatedly discovered not to have been signed by the payee but by his wife, who cashed them when she went shopping.[13])

Standards are of two basic types, "request standards" and "collected standards." The former have the advantage of producing the desired wording and letter combinations, but they also have serious drawbacks. One is that they may lack the writer's natural handwriting flow; another is that the suspect may deliberately disguise his or her handwriting. Both problems can, however, be partially prevented by having the material dictated rather than copied and by having it repeated several times. The collected (or "nonrequest") standards are specimens gathered from any of various sources. Canceled checks are a readily available source of dated signatures. Standards of other writing may come from letters, manuscripts, diaries, and the like. Both types of standards—request and collected—can be used together in a case. Request standards can supplement collected ones that are in insufficient quantity, and collected standards can be compared with request exemplars to ensure against disguised writing.[14]

Identification of handwriting is based on a number of factors which may be grouped into three main categories: form, line quality, and arrangement.[15]

Form. This refers to a number of characteristics related to the shape of the elements in the handwriting (or handprinting). One of these is the *formation of letters*, the shape of the individual upper- and lowercase letters and numbers. Naturally, the examiner looks for distinct formations that vary from the copybook models. The letter formation may also to some extent be a function of the *movement* of the writing, whether produced by the fingers (usually lacking smooth lines and having broad curves), by hand movement (a somewhat freer writing), and forearm movement (which allows the ultimate in freedom and may be typified by smoothness in the long strokes). Other important aspects of form include *proportion* (the relative height of letters), *slant* (when the axes of certain letters consistently deviate from the general slant of the writing), *retrace* (when the staffs of certain letters such as the *t* are retraced rather than looped), various *angles, straight lines, and curves* (wherever they depart from the norm), *connections* (the strokes that link one letter with another), and what are termed *trademarks* (very distinctive noncopybook features such as a plus sign used for "and" or the filling in of the space on a check between the written amount and the printed word "dollars" with a wavy line or a series of dots).

Line quality. This group of characteristics may be influenced by the type of *writing instrument* used (the flexible nib pen producing lines that are more expressive and revealing of a writer's habits than the ball-point or pencil). The nib tends to produce accented lines on the downstroke called *shading*, a careful analysis of which reveals the writer's habitual *pen position* as well as the amount of *pressure* exerted. A flowing script has a smooth line that indicates *speed and rhythm*, whereas a slowly, laboriously penned script may be characterized by uniform pressure and even shakiness or *tremor*. *Beginning and ending strokes* may also indicate the speed of writing (usually being tapered in the case of rapid writing because the pen is in motion, or blunt from the pen being placed on the paper before the writing commences). Line quality may also include *continuity* (characterized by connecting strokes between letters).

Arrangement. The final category of handwriting identification factors consists of those related to placement. Among these are *spacing* (the distance between letters, words, and lines) and *alignment* (the relationship of the writing to the "base line," an actual or imaginary line on which the handwriting rests). Evidentiary value may also be found in the width of margins and their vertical alignment or lack thereof. So may other aspects of *formatting*, which refers to the layout or arrangement of elements, such as the parts of a letter, on a page. Related to formatting is signature *placement* (its positioning relative to the body of writing). Other arrangement factors include distinctive *punctuation* and *corrections*, which may occur in many forms. (This list is not exhaustive. At least one source lists *spelling* as an identification trait, although we consider it as a form of "internal evidence," which is discussed in the next section.)

The identification factors are used to compare the questioned writing with the known standards. First, however, the disputed or questioned writing is examined to determine that it is naturally and freely executed so that there are therefore unconscious, habitual movements, and that the writing was not drawn, traced, or in some manner forged.[16] The questioned writing is examined first without reference to the standards, so as not to be "influenced by the pictorial resemblance which often exists between two handwritings." In this way, the examiner's judgment remains unbiased.[17]

The second step is to examine carefully the known or standard writings to determine what the genuine handwriting habits of the individual are—that is, the habits of form, line quality, and arrangement. (Hand printing relies on somewhat different characteristics, but the basic procedure is the same.)

Finally, the examiner compares the questioned writing with the standards. As O'Hara explains: "In comparing two specimens of handwriting the expert searches for characteristics which are common to both the questioned and standard writing. If the characteristics are sufficient in kind and number and there are no significant unexplainable differences, he may conclude that the writings were made by the same person."[18] It has been estimated that a handwriting specimen might have between five hundred and one thousand individual characteristics. One police-science text states: "The theory upon which the document expert proceeds is that every time a person writes he automatically and subconsciously stamps his individuality in his writing," and therefore "through a careful analysis and interpretation of the individual and class characteristics, it is usually possible to determine whether the questioned document and the standards were written by the same person."[19]

Obviously, some characteristics will be more distinctive than others. Some will be so unique—like those of Bruno Hauptmann—that they will carry great evidential value. In any case, the modern tendency is to avoid attempting to tabulate the individual characteristics for each of the two writings; that is, to avoid attempting precise calculations except in special cases. Instead, the expert uses his knowledge and experience to insure that, as Ordway Hilton states, "the same distinctive, personal writing characteristics are found in both the known and unknown writing in sufficient number and that the likelihood of accidental coincidence is eliminated—and that there are no basic or fundamental differences between the two sets of writing."[20] He continues:

> The document examiner is occasionally asked how many points of identification are necessary to establish that two writings are by the same person. Such criteria have not been established, and probably could not be, because of the nature of handwriting identification. It involves not only factors of form that are subject to relatively easy count, but also the qualities of execution, freedom, movement, skill, emphasis, spacing, and the like that influence the entire writing and are not susceptible to tabulation. As a consequence, the combination of a unique set of similarities coupled with the lack of significant basic writing differences must be used as the true basis for a positive identification.[21]

The FBI laboratory has long maintained handwriting reference files, including an Anonymous Letter File, the National Fraudulent-Check File, and the Bank Robbery Note File. Also, today there are sophisticated, automated techniques for the identification of signatures and other handwriting. For example, the Forensic Information System for Handwriting (FISH) permits forensic document experts to scan and digitize such writings as ransom letters or threatening correspondence. These images may then be compared to images in the FISH database, and a list of the most probable "hits" produced. The questioned handwritings, along with the nearest hits, are submitted to a questioned-document examiner for a final determination.[22]

We should say a word about disguised writing and printing. The most common disguising ploy is to change the direction of slant. (Other means include changing the writing hand, writing upside down, altering the size of the writing, and substituting hand printing for cursive writing.) Despite such techniques, writing patterns tend to be so habitual that they are difficult to suppress or alter. It is not unusual for two writings to match, point for point, except for—say—a backhand slant imparted to the questioned writing in a futile attempt to disguise it. The attempt to disguise is apt to be more successful in a short writing than a long one. As Hilton observes, "The task of maintaining an effective disguise grows more difficult with each additional word."[23]

As with handwriting, typewriting has both class and individual characteristics. Class characteristics are those of a particular make and model of machine, identifiable by the specialist examiner from the typeface by comparing it with a reference collection that may consist of thousands of type specimens. Certain points of comparison are useful in determining whether two specimens of typewriting were produced by the same model of typewriter: the overall size of letters; the lengths of serifs, or the horizontal bars at the terminals of strokes; the relative curvature of endings (in *f, g, m, t,* and *y*); and the size and design of figures (for example, the relative areas and shapes of the ovals in the *8*).[24]

The individual characteristics are those that develop through use and abuse, such as wear and faulty alignment. The specific faults depend on the type of machine—the common old shift-key typewriter, the IBM "Selectric" (introduced in 1961), or the modern word processor. With the latter, the printer's type wheel is the most likely source of identification characteristics. The typefaces can become worn, and slight alignment and printing defects can also occur. Because of the complexity of modern typewriters and word processors, some document examiners have become specialists in this narrow field. As with handwriting and handprinting, comparisons of typewriting begin with the acquisition of suitable standards or exemplars, which should be as similar as possible to the questioned writing in terms of paper and cleanliness of typeface. Standards should be used that were made as near the date of the questioned typewriting as possible.[25]

FORGERY DETECTION

Apart from handwriting comparison, forgery detection represents the major portion of the work of the document specialist. In this section we examine the forger's techniques and the warning signs that point to forgery. In attempting to fraudulently reproduce a particular handwriting, such as a given person's signature, the forger resorts to one of a few methods: tracing, freehand copying, or mechanical placement.

Tracing is the most amateurish means of forging a signature or (usually brief) text. Typically one of two means is employed: the trace-over method or the light box technique. In the trace-over method, the faint outline of a genuine signature is transferred onto a sheet of paper placed underneath it by means of heavy pressure or the use of transfer paper. This outline, either an indented or a graphite- or carbon-paper copy, is then traced over in ink with an appropriate pen. An obvious drawback of such an approach is that it tends to leave evidence. It is difficult to follow the outline exactly, so traces of the indentations or the carbon or graphite outline may show in the final pen work. And although the graphite traces may be erased, the erasure itself may be detected, as discussed in the following section on analytical techniques.

Similar problems occur with the second method of tracing—using a light box or window. In this method, the original signature is placed under the sheet used for the forgery, while backlighting renders the writing visible through the overlying paper. This method of tracing leaves no telltale traces on the forged document. Nevertheless, the result will usually have a belabored appearance or at least will lack the smooth quality of natural penmanship, and, even with very thin paper and strong lighting, some of the fine detail of the writing will inevitably be lost.[26]

Traced forgeries also may be detected when the suspected model for the forgery is available. In this case, the tracing will closely superimpose over the model signature or writing, whereas with genuine signatures no two are ever exactly the same and so are unlikely to closely correspond under superimposition. A famous case involving four signatures, each on a different page of a legal instrument, was that of the W.M. Rice will, a document dated June 30, 1900, that was designed to defraud the estate, which was valued at more than $6 million. Genuine signatures of Rice made on the same day showed the natural range of variations, but the signatures on the will were unnaturally similar, almost as if they had been produced by a rubber stamp. (They lacked other features of genuine writing as well—notably the natural shading of pen strokes.)[27]

Far superior to tracing is the freehand technique of producing forgeries—at least in theory; a good tracing may still be better than an ineptly drawn one. The most inept of freehand forgeries is the "spurious signature," that is, a made-up signature in the forger's own or a disguised hand.[28] Although somewhat more successful, the slowly copied forgery is produced in a manner similar to tracing and therefore often has similarly poor qualities.[29]

Much more successful is the practiced freehand forgery. A talented artist or calligrapher who has taken the time to practice a given signature may eventually learn to sign a name or, with considerably more difficulty, imitate handwriting that is remarkably similar to the targeted writing and that is comparatively smoothly and freely written.[30] Still, the forger's own idiosyncratic traits tend to creep into the writing, and there are other problems that the document examiner may discover. For example, a forger who produced fake Robert Frost manuscript poems that were excellent productions, nevertheless omitted the personal inscriptions Frost invariably penned below the poems he copied for his admirers.[31]

In addition to tracing and freehand copying, there are methods of mechanically placing a signature onto a check: projections, stampings, and signature splitting. Projections involve using an optical system (xerography) to reproduce a given signature onto a check or other document. There also are special devices used by security agencies for producing clandestine documents. Stampings involve the simple forging of rubber-stamped or imprinted facsimile signatures, which are increasingly being utilized on corporate checks. A "split" signature is one that is lifted off a genuine signature with transparent tape and then transferred to a forged document—usually a check that is made to appear torn at that point.[32]

Among the numerous indicators that a writing may be spurious are the following: evidence of tracing or prior drawing, forger's tremor (shaky handwriting, usually coupled with other signs of forgery), evenness in pen pressure, unnatural hesitations, uncertainty of movement, blunt beginnings and endings, unnatural pen lifts (made by the forger to check the progress of his work), patching (careful retouching), uncommon forms (anything that differs from a writer's usual habit), off-scale writing (writing much larger or smaller than the writer's usual script), or excessive attention to detail.[33] The following nonhandwriting factors also may indicate forgery: incorrect writing materials, lack of provenance (or historical record—in the case of historical documents), or internal evidence (mistakes of format, spelling, content, etc.)[34] In addition, various scientific analyses, which are discussed in the following section, also may reveal the forger's handiwork, as well as restore erased writing and assist the document expert in the various tasks that come to him or her.

ANALYTICAL TECHNIQUES

The basic techniques for analyzing questioned documents include macroscopic and microscopic study, various spectral techniques, and certain chemical and instrumental tests.

Macroscopy (as distinct from microscopy) is the scrutiny of things visible to the naked eye or with an ordinary magnifying glass. It may be conducted by reflected light, oblique light, or transmitted light. The usual viewing of a document, in which light falls normally on the viewing surface, is known as reflected-light examination. This basic inspection technique may reveal such fundamental errors as incorrect paper or pen choice or errors of format.[35]

Oblique-light examination is another macroscopic technique with considerable potential. Also called side-light or grazing-light examination, it is conducted with light striking the document's surface from one side at a low angle. This technique reveals the shadows that are produced by any surface irregularities, notably erasures (the roughened surface of which may be revealed), indentations (as from traced writing or writing traces inadvertently left on the topmost sheet of a pad of paper), embossments (such as embossed seals), and the like. Writing or typewriting on a charred document may also be deciphered in this way. The sheen of the ink relative to that of the blackened paper may provide suitable contrast under oblique lighting so that the writing may be read and photographed.[36]

The examination of a document by transmitted light involves illuminating it from behind, as by placing it on a light box, so that the light passes through the paper. Such an examination facilitates identifying the type of paper, studying any watermarks that might be present, and detecting erasures or other alterations.[37] Changes in watermarks can provide a means of dating a document. For example, a will dated 1912 was purportedly discovered in an old trunk by a relative of the deceased man. However, the document examiner discovered that the watermark spelled out "erkshire Bond." Subsequent checking with the Berkshire Company revealed that the paper was not made until 1916 and that sometime between 1917 and 1918 the *B* had fallen off the special roller used in imprinting the watermark. Therefore, the will could not have been written in 1912 but was a forgery produced at least five years later.[38]

As with watermarks, transmitted light also reveals thin spots in the paper caused by erasures (especially in the case of the abrasive, gray "sand rubber" type of eraser designed to remove ink writing). Backlighting also reveals the opposite effect—increased opacity rather than translucency. This can be caused by application of various types of "correction" materials such as the "white out" that comes in various formulas for pen and ink, photocopies, etc. In fact, text typed or written under such corrected areas can often be read by transmitted light.[39]

In contrast to macroscopy, microscopy is used to provide higher magnification in document work. A relatively low-powered stereoscopic or

FIGURE 7.2. Indented writing—as is often found on the top page of a notepad—may be enhanced by means of light directed at an oblique angle to the page. The enhanced writing may then be photographed.

"stereo" microscope generally is used for direct inspection of a document. (Usually such a microscope is in the 10- to 60-power range; 20- to 30-power is especially useful for most document work.) The stereomicroscope is typically mounted on an extension arm that permits use over large documents. The stereoscopic feature provides a high-resolution, three-dimensional image, enabling the document examiner to view more accurately such subtle, depth-related features as nib tracks (furrows in paper left by steel or other hard-nibbed pens), crossed pen strokes, and erasures. The stereomicroscope is ideal for determining the type of pen used to produce a given writing, as well as for detecting and studying "patching" (retouching) of writing, pen lifts, tremor, erasures, corrections or other alterations, sequence of pen strokes where one crosses another (usually, in the case of fluid inks, the second line tending to spread into the first at the intersection), the identifying features in typewriting, and many, many additional elements.[40]

The standard laboratory microscope is reserved for examining minute traces, such as a tiny sample of ink or paper removed from a document. By using a "well" slide (one with a central recess), microchemical tests can be conducted on such samples, with the reagents (or test chemicals) being applied with an eyedropper or a hypodermic syringe, and the reaction being observed under the microscope. Such microscopes typically have three or four objective lenses, offering a range of powers such as 40x, 100x, and 400x.

Both macroscopic and microscopic features can be photographed. For the former, photomacrographs are made with a good quality copy camera, one that is able to photograph details from 1-to-1 to 10-to-1 magnification and that is capable of photographing either a whole document or any portion of it. This camera can be used in conjunction with flanking lamps and an ordinary photographic copy stand that secures the camera while permitting it to be moved up or down. Oblique lighting is accomplished by illuminating only one of the flanking lamps, and transmitted light is derived from a light box that can be set on the copy stand. In addition to photomacrographs, photomicrographs (photos made through a microscope) may be made. Photographs enable the examiner to document observations and utilize them as evidence in the courtroom.[41] Document experts also have at their disposal various "spectral" techniques—those that take advantage of the invisible as well as the visible spectrum. These include ultraviolet light, infrared radiation, laser technology, and special photographic processes.

Although ultraviolet light is invisible, its effects on an object as observed in a dark room may be distinctly visible; the object is then said to *fluoresce.* This interesting phenomenon is useful to the investigator who may in this manner detect alterations on a check, secret writing in a letter, or the presence of optical brighteners in paper. (The presence of such brighteners in the "Hitler Diaries" was an early indicator that the diaries were not genuine, since such whitening agents were not used prior to the 1950s.[42]) Other uses include detecting various erasures and corrections, enhancing faint or erased writing, revealing dissimilar glues (in the case of an envelope that has been opened and resealed), evidence of repair in the case of a tampered wax seal, certain signs of age in a document, and other effects that may be invisible to the unaided eye but quite apparent under ultraviolet light.[43]

Infrared illumination is at the opposite end of the visible spectrum, lying between visible light and radio waves. Its effects must be viewed by special optical means—either a special viewing device or photography with special filter and film. Despite its practical limitations, infrared illumination offers a panoply of remarkable investigative possibilities. Art experts use it to detect undersketching in paintings and alterations in

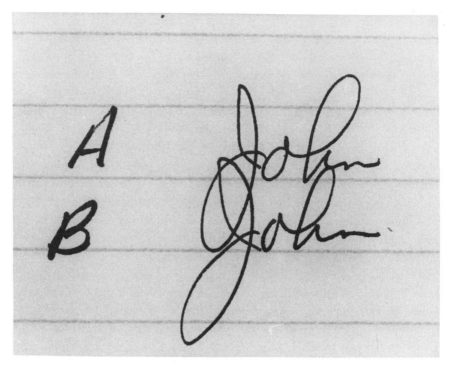

FIGURES 7.3 AND 7.4. Different inks may be indistinguishable to the unaided eye (fig. 7.3) but appear quite different when photographed using infrared luminescence photography (fig. 7.4).

artworks of various kinds. Infrared is used by forensic experts to decipher charred documents, to develop erased writing, and to differentiate between certain types of ink. (Some inks *absorb* infrared radiation and thus darken; others *reflect* the rays and lighten, while still others *transmit* the infrared and thus disappear.)

By this means it may be possible to distinguish between two different inks used for a document, even though the inks appear the same by ordinary observation. Falsified checks that are "raised" by using some cleverly added pen strokes, to alter, say, the "5.00" and "five dollars" to read "50.00" and "fifty dollars" may be detected if the original ink and the ink used by the forger are of different types with regard to infrared response. So may similar alterations be detected, such as a postal-meter stamping with a retouched, raised figure for the amount.[44] Infrared radiation also may be used to detect and read secret writing, read unopened letters (when the paper is transparent to infrared and the ink opaque to it), detect and differentiate stains, and perform many other investigative tasks.[45]

FIGURE 7.4.

Laser technology is a more recent weapon in the fight against forgery and other criminal acts. In chapter 5 we noted its importance in detecting fingerprints. In document work it may be found useful in many of the areas previously limited to ultraviolet and infrared examinations and may prove successful when the other techniques are not. For example, with reference to inks, sophisticated laser technology has been utilized to discriminate between similar inks when other, conventional methods have proven unsuccessful. This involves using what is known as laser-luminescence spectroscopy by which it is possible to analyze very small amounts of ink *in situ* (in place on the document). As a less expensive alternative to laser technology, specially filtered high-intensity light is increasingly being utilized in forensic work.[46]

Among special photographic techniques employed in questioned-document examination, there are naturally ultraviolet, infrared, and laser photography. In one case investigated by the FBI, ultraviolet photography rendered legible the eradicated writing on a Tennessee fraudulent automobile registration certificate that had been used to sell a stolen vehicle. Similarly, infrared photography was instrumental in restoring part of an erased letter in an altered postmark (in which "JUN" was changed to "JAN"). Even conventional photography offers many useful techniques such as increasing or decreasing contrast and using any of a variety of available filters.[47]

Thus far we have discussed only nondestructive tests. While these are essential for valuable or historical documents, the nature of forensic work is such that somewhat destructive techniques (such as removing a small sample of paper from a document) also must be utilized. Among the chemical tests used by document experts are those applied to paper and ink. The current standard forensic procedure for identifying inks is by a process called chromatography, usually thin-layer chromatography. In this process, very tiny plugs are punched from the inked areas in question, and the ink is then extracted by means of a solvent. The dissolved ink is then spotted onto a silica-gel chromogram sheet, which is placed in a developing tank for an interval of time and then allowed to dry. The resulting chromatographs of the questioned inks are compared with those of standard inks of the same type and color.[48] The most complete collection of writing ink reference samples is the seven-thousand-specimen International Ink Library housed in the Questioned Document Branch of the U.S. Secret Service Forensic Services Division.

Other chemical methods may reveal certain hidden features in a document. For example, there are chemical solutions that may intensify ordinary pencil traces remaining from an erasure. Also, iodine fuming may enhance indented writing and reveal chemical and other erasures as well as develop latent fingerprints. (The developed traces usually fade, so they should be photographed promptly.)[49]

As a field, questioned-document examination is increasingly coming under attack as a "black art"—in part because of the degree of subjectivity that may be involved. The field does have the unfortunate distinction of attracting more conflicting testimony than any other crime-laboratory discipline. In response, however, a number of corrective procedures have either been implemented or recommended, including eliminating graphology and other pseudosciences from the courtroom, requiring appropriate training and board certification of examiners, reducing heavy caseloads, ensuring that sufficient and appropriate exemplars are obtained for comparisons of handwriting, and urging ethics committees to deal with unethical examiners; i.e., those willing to resolve any issue according to the wishes of the client. These and other reforms may help ensure that questioned-document examination meets the fundamental

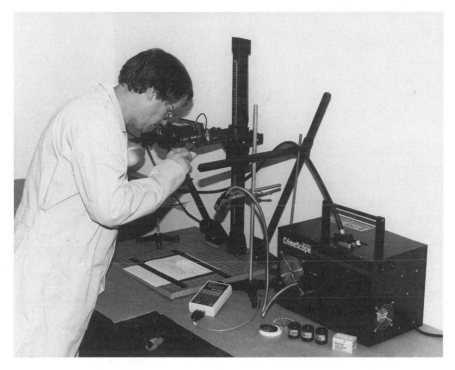

FIGURE 7.5. Infrared photography is often used to distinguish between visually similar types of inks. Here co-author John F. Fischer uses the infrared output of the "CrimeScope" and a Nikon camera equipped with an infrared lens and filter to examine the ink on a document.

criterion of reproducibility of results and thereby receives the credibility it deserves.[50]

As we shall see in the following case study, a variety of techniques may be needed in a given case—even techniques that sometimes must be innovated to solve particular problems.

CASE STUDY: THE MORMON FORGERY MURDERS

Salt Lake City, October 15, 1985: Within two hours' time, first one and then another pipe bomb exploded, destroying the calm of this largely Mormon city and sending shock waves across the country. Dead were Salt Lake businessman Steve Christensen and a suburban grandmother, Kathy Sheets. The following day, just as panic was beginning to set in, a third bomb exploded inside the car of Mark Hofmann, a young and well-respected dealer in historical documents. The blast ripped the roof off Hofmann's blue sports car, but he survived—minus a kneecap, bleeding from various additional wounds, and blackened from powder burns.[51]

Questioned at Latter-Day Saints Hospital by police, Hofmann seemed unaccountably *evasive*. And his explanation of finding a strange package in his car and reaching for it when it blew up was untrue. Bomb experts knew that Hofmann was actually "kneeling on the seat fiddling around with the bomb when it went off."[52] Soon the number-three victim became the number-one suspect. It was theorized that Hofmann was taking a bomb to a third victim when it exploded accidentally. Hofmann even fit the description of a man seen carrying a package to Christensen's office, and a green jacket like the one the man had worn was discovered during a search of Hofmann's bungalow.[53]

Yet, although his lie-detector test was inconclusive, as were two more such tests required by Utah law, a fourth showed that Hofmann was telling the truth.[54] He was apparently a devout Mormon as well as a document dealer with a sterling reputation. It was unthinkable, said his many friends and his associates in the world of rare manuscripts, that Hofmann could be guilty of murder. The same was said as evidence began to suggest that rare documents recently sold or offered for sale by the mild-mannered and personable young dealer were forgeries. If true, this suggested a possible motive for the murders: to cover up a high-stakes fraud.

One of Hofmann's sensational document "finds" was what is known as the "Oath of a Freeman." Although the seventeenth-century oath is known to historians as the first example of printing in America, no actual specimen of its printed text had ever come to light. Then in the mid-1980s first one and then another copy surfaced, both allegedly discovered by Mark Hofmann—who seemed to have a talent for finding what others could not. His discoveries included sensational Mormon papers such as the controversial "white salamander letter"—a missive purportedly written by an early Mormon leader that undermined some of the church's major tenets. Among Hofmann's other offerings were rare historical documents and currency, including letters and other autobiographic material from such historical figures as Daniel Boone, mountain man James Bridger, Betsy Ross, Charles Dickens, and Mark Twain.

Many of Hofmann's suspected forgeries had already been authenticated by some of America's premier autograph and manuscript experts. And the salamander letter had been all but authenticated by the FBI laboratory, which, while conceding that a "lack of sufficient known signatures and writing [i.e., standards] prevented a definite conclusion," nevertheless stated in its report: "These writings appear to have been normally written and no evidence was observed which would indicate forgery or an attempt to copy or simulate the writing of another. . . . There is no evidence to suggest that these documents were prepared at a time other than their reported dates."[55] Unfortunately the FBI lab had relatively little expertise in historical writings, and the manuscript experts who had authenticated some of Hofmann's documents often had scarcely looked at them but merely had relied on his excellent reputation.

Now, however, the Salt Lake City County attorney's office was prosecuting Hofmann, and eventually several of the historical documents were examined by forensic experts who had been extensively assisted by manuscript experts. (For instance, the eminent dealer Charles Hamilton of New York was retained as a consultant and, on his personal recommendation, one of us [J.N] was consulted by a prosecutor, Gerry D'Elia, about the making—and artificial aging—of iron-gallotannate ink, the standard ink for documents until the twentieth century.) This proved an effective combination of experts with various areas of expertise and ultimately resulted in detection of forgeries. Based on their often-suspicious provenance, their incredible rarity, and the sensational content of some of the papers, the Hofmann documents deserved the closest scrutiny.

It was, in fact, manuscript experts who first branded the "Oath of a Freeman" a fake. This came after one scientist concluded that the ink's bonding to the paper was consistent with an age of some three hundred years—supposedly proof that the document was indeed authentic. Suspicions were heightened, however, by Hofmann's "discovery" of a *second* copy of the oath, a document-world equivalent of lightning striking twice. As it happened, Hofmann had used an artificial aging technique to reproduce the bonding effect of old printing ink. Certain typographical anomalies were fatal, however, enabling scholars to demonstrate the document's spuriousness despite scientific pronouncements to the contrary. For example, printing experts observed that there was an overlapping of descenders (the tails of letters like *j*) with ascenders (letters like *b* and *d*) in the succeeding lines. Such overlapping could never happen with authentic hand-set type wherein each line is self-contained.

Hofmann's approach to the forgery had been nothing less than brilliant. Since the "Oath of a Freeman" had been printed by Stephen Daye, who had also printed the *Bay Psalm Book*, Hofmann obtained a cheap facsimile of that volume and cut out the requisite letters, rearranging them into the text of the "Oath" (the wording surviving from later editions of the long-lost original). From this pasted-up model he had a printing plate made. He gave further details of his craftsmanship to the prosecutors as part of a plea bargain through which he escaped the death penalty. He obtained real seventeenth-century paper from the flyleaves of books in the Brigham Young University Library. He burned one such sheet to obtain carbon black for his ink—a precaution in case the ink was carbon-14 tested—and he mixed that with boiled linseed oil, tannic acid (obtained by boiling a piece of seventeenth-century leather binding in distilled water), and a bit of beeswax. He "aged" the new printing plate by grinding a few letters and rubbing the whole with steel wool. He rolled ink onto the plate, then covered it with felt and another metal plate, and pressed the affair with a C-clamp in his basement workroom. To give the oath a semblance of provenance, Hofmann claimed that he had bought the document, and he cleverly counterfeited proof thereof. He simply printed a new title, "Oathe of a Freeman," on an inexpensive Civil War broadside, added a price of "$25" in pencil, and then mixed it with other inexpensive items kept in a browser's folder in the Argosy Book Store in New York City. There he "discovered" it and dutifully paid a clerk the twenty-five-dollar price. Hofmann asked the clerk for a receipt and was given one that proved he had purchased an "Oath of a Freeman" at Argosy on a given date.[56]

Hofmann took a similar approach to his other forgeries, building on his own background as a document dealer. He did historical research, obtained antique paper, made ink from old recipes, cut quill pens, took a course in calligraphy and practiced early penmanship, and conducted experiments in artificially "aging" ink on the paper—even applying suction from a vacuum cleaner to the back of a sheet (supported by screen wire) to draw a chemical oxidant deep into the paper and thus simulate the effects of time. He also used such special techniques as heating with a hand iron or applying chemical oxidants such as peroxide to artificially age the writing (converting the fresh black iron-based ink to a rusty brown oxidized state).[57]

Forensic experts George J. Throckmorton and William Flynn did discover signs of the artificial aging. These included the presence of ridges that had the appearance of scorching with an iron, a bluish discoloration under ultraviolet light (caused by the chemical treatment) coupled with a unidirectional ink migration, evidence that the document had been treated with liquid chemicals and hung up to dry. (Genuinely old ink would naturally migrate outwardly in all directions over time.) The experts also discovered a property of the Hofmann documents that differed from the genuine historical papers that they used for standards:

Hofmann's documents had ink that exhibited an "alligatoring effect," a cracking that was different from the real effects of age, caused by the sudden, extreme effects of the chemical oxidants. And microscopic examination showed that, when the crusted ink was scraped away with a sharply pointed instrument, the paper beneath was not stained brown as it would be in genuine old documents.[58]

There were other problems, one of which manuscript experts had pointed out regarding the white salamander letter—it had been folded and sealed improperly for a pre-envelope letter dated 1830. Also, since pre-envelope covers also bore postmarks, Hofmann had faked one on the address panel of the salamander letter, but, as an expert reported, "the flatness, vagueness, and ink distribution of the postmark differ from genuine postmarks of the period."[59] In one or another document, flaws were detected in the printing, letter paper was discovered to have been purloined from old books, linguistic analysis indicated that the text of one letter was authored by someone other than its putative author, and so on. Even earlier, of course, circumstantial evidence had begun to point toward Hofmann, and a search warrant uncovered incriminating evidence in his possession, such as a check made out to the printing firm that made the plate for "Oath of a Freeman." (The plate had a disclaimer at the bottom that had allayed the printer's fears, but Hofmann omitted the disclaimer when he printed his fake version.)[60] The Hofmann case brings to mind the words of noted manuscript expert Mary Benjamin in her *Autographs* (1986). As she wisely comments:

> To differentiate the natural handwriting of an individual from that of another is not too difficult a task for the adept, but to detect forgeries is a very different matter, requiring greater alertness, patience, study, and skill. The professional expert, for instance, has at his disposal fairly well perfected, modern and scientific devices, such as measuring instruments, light rays and chemical tests with which he can make a thorough analysis of all materials. The dealer-expert, on the other hand, is equipped with complementary advantage of long experience. In addition to a subconscious guiding instinct, he draws on a heterogeneous fund of information. Generally he possesses such a photographic memory that without ever seeing the signature he can recognize at a glance the handwriting of hundreds of famous men and women. He is, moreover, familiar with those personal affectations which led them to select a particular type, color and size of paper, a particular kind of ink or a thick or thin pen. He knows certain eccentricities which distinguish an individual's script—the size of strokes, how letters are looped, how "t's" are crossed and "r's" formed, how words are spaced and many

other revelatory features. This is a special knowledge gained by years of handling thousands of miscellaneous letters, which even the professional expert does not have. That each one can happily supplement the work of the other is obvious, and on many occasions they have pooled their resources.[61]

Sentenced to life imprisonment on January 24, 1987, Mark Hofmann entered the Utah State Prison at Point of the Mountain. He soon wrote the following letter to the family of murder victim Kathy Sheets:

> Of course it is difficult to explain my actions of October 1985. I cannot justify what occurred. My actions have caused irreperable harm to your families and to my own.
>
> Saying I'm sorry sounds so hollow as to seem meaningless, but with all my heart I want you to know that although my actions were inexcusable, I am sorry.
>
> I have tried in small measure to rectify my crimes by fully exposing them to the authorities, and putting to a halt the further trama of the trial process.[62]

Mark Hofmann's legacy is to remind us that forgers will go to tremendous lengths to accomplish their nefarious tasks, that questioned-document examiners must therefore remain ever watchful for innovative new techniques, and that a comprehensive or "multi-evidential" approach—one that considers handwriting, provenance, linguistics and other internal evidence, evidence from writing materials, and the results of scientific analyses—is essential.[63]

NOTES

1. Herbert Danby, *The Mishnah*, translated from the Hebrew (Oxford: Oxford Univ. Press, 1933), 247.

2. See the two chapters by Joe Nickell in *The Write Stuff: Evaluations of Graphology*, ed. Barry L. Beyerstein and Dale F. Beyerstein (Buffalo: Prometheus Books, 1992), 23-29, 42-52. For an abridged version, see "Graphology Versus Science," in Joe Nickell, *Detecting Forgery: Forensic Investigation of Documents* (Lexington: Univ. Press of Kentucky, 1996), 17-20. (This chapter is abridged from the latter source.)

3. Wilson R. Harrison, *Suspect Documents: Their Scientific Examination* (New York: Frederick A. Praeger, 1958), 518-19.

4. Nickell, *Detecting Forgery*, 20-21.

5. Stuart Kind, "Criminal Identification," in *Science against Crime*, ed. Yvonne Deutch (New York: Exeter Books, 1982), 28.

6. Ordway Hilton, *Scientific Examination of Questioned Documents,* rev. ed. (New York: Elsevier Science, 1982), caption to illustration, 166.

7. David Fisher, *Hard Evidence* (New York: Dell, 1995), 242.

8. Nickell, *Detecting Forgery,* 21-22.

9. "Identification of Handwriting" forensic instruction manual for course in scientific crime detection (Chicago: Institute of Applied Science, 1962), lesson 1, 5.

10. Nickell, *Detecting Forgery,* 26.

11. Albert S. Osborn, *Questioned Documents* (1910; 2nd ed. Montclair, N.J.: Patterson Smith, 1978), 249.

12. E. Patrick McGuire, *The Forgers* (Bernardsville, N.J.: Padric Publishing, 1969), 208.

13. Hilton, *Scientific Examination,* 309; photograph, 310.

14. Nickell, *Detecting Forgery,* 29-35.

15. These three categories are condensed from four offered by Charles E. O'Hara in *Fundamentals of Criminal Investigation,* 3d ed. (Springfield, Ill.: Charles C. Thomas, 1973), 786; and rearranged, with some borrowing from "Identification of Handwriting" (lessons 1-3), and Hilton (*Scientific Examination,* 154-60).

16. Gideon Epstein, based on a quotation in Nickell, *Detecting Forgery,* 43.

17. "Identification of Handwriting," lesson 4, 1.

18. O'Hara, *Fundamentals,* 785-86.

19. Fred E. Inbau, Andre A. Moenssens, and Louis R. Vitullo, *Scientific Police Investigation* (New York: Chilton, 1972), 50.

20. Hilton, *Scientific Examination,* 161.

21. Ibid.

22. "FSD," current brochure of Forensic Services Division, Dept. of the Treasury, U.S. Secret Service, n.d.

23. Hilton, *Scientific Examination,* 169. See also Nickell, *Detecting Forgery,* 48-50; O'Hara, *Fundamentals,* 795-97, and Osborn, *Questioned Documents,* 410-11.

24. O'Hara, *Fundamentals,* 801-2.

25. Inbau, Moenssens, and Vitullo, *Scientific Police Investigation,* 64-65; Hilton, *Scientific Examination,* 334-35.

26. Nickell, *Detecting Forgery,* 59-60.

27. Osborn, *Questioned Documents,* 329-31.

28. Hilton, *Scientific Examination,* 191.

29. Osborn, *Questioned Documents,* 115.

30. Nickell, *Detecting Forgery,* 61.

31. Charles Hamilton, *Great Forgers and Famous Fakes* (New York: Crown, 1980), 179.

32. McGuire, *The Forgers,* 173-75.

33. Nickell, *Detecting Forgery,* 59-92.

34. See ibid., 95-126, for an extensive discussion.

35. Ibid., 127-32.

36. Ibid., 132-35.

37. Ibid., 137-39.

38. Henry Morton Robinson, *Science Catches the Criminal* (New York: Blue Ribbon, 1935), 112-13.

39. Nickell, *Detecting Forgery*, 142-44.

40. Ibid., 145-51.

41. Ibid., 151-53.

42. Charles Hamilton, *The Hitler Diaries* (Lexington: Univ. Press of Kentucky, 1991), 102.

43. Nickell, *Detecting Forgery*, 154-59.

44. Ibid., 160-66.

45. O'Hara, *Fundamentals*, 766-67.

46. Nickell, *Detecting Forgery*, 166-69.

47. Ibid., 170-76.

48. Richard L. Brunelle, "Questioned Document Examination" in *Forensic Science Handbook*, ed. Richard Saferstein (Englewood Cliffs, N.J.: Prentice-Hall, 1982), 712-15; Nickell, *Detecting Forgery*, 178-85.

49. Ibid., 185-95.

50. Floyd I. Whiting, "The Questioned Document Expert: Skilled Specialist or Pseudoscientist," *International Journal of Forensic Document Examiners,* vol. 2, no. 4 (Oct./Dec. 1996): 284-87.

51. Steven Naifeh and Gregory White Smith, *The Mormon Murders: A True Story of Greed, Forgery, Deceit, and Death* (New York: Weidenfeld & Nicholson, 1988), 3-36. Except as noted, this narrative is taken from this source.

52. Ibid., 48.

53. Ibid., 45-51.

54. Ibid., 319-21.

55. FBI report, quoted in Charles Hamilton, *Great Forgers and Famous Fakes,* 2nd ed. (Lakewood, Colo.: Glenbridge Publishing, 1996), 282.

56. Naifeh and Smith, *Mormon Murders,* 435-37; Hamilton, *Great Forgers,* 287-88.

57. George J. Throckmorton, "A Forensic Analysis of Twenty-One Hofmann Documents," appendix of Linda Sillitoe and Allen D. Roberts, *Salamander: The Story of the Mormon Forgery Murders* (Salt Lake City: Signature Books, 1988), 531-52.

58. Ibid.

59. Ibid., 544.

60. Sillitoe and Roberts, *Salamander,* passim; Naifeh and Smith, *Mormon Murders,* 380-81.

61. Mary Benjamin, *Autographs* (New York: Dover, 1986), 143.

62. Quoted in Naifeh and Smith, *Mormon Murders,* 440.

63. Nickell, *Detecting Forgery,* 95-98.

RECOMMENDED READING

Hamilton, Charles. *Great Forgers and Famous Fakes: The Manuscript Forgers of America and How They Duped the Experts,* 2nd ed. Lakewood, Colo.:

Glenbridge Publishing, 1996. An essential text for anyone interested in historic forgeries, providing an entertaining look at the lives of forgers; nicely illustrated.

Hilton, Ordway. *Scientific Examination of Questioned Documents.* Rev. ed. Amsterdam: Elsevier, 1982. An excellent modern text on forensic document examination.

McGuire, E. Patrick. *The Forgers.* Bernardsville, N.J.: Padric Publishing, 1969. A wide-ranging study of the crime of forgery, including modus operandi, case histories, personality profiles, and forgery prevention.

Nickell, Joe. *Detecting Forgery: Forensic Investigation of Documents.* Lexington: Univ. Press of Kentucky, 1996. A "multi-evidential" approach to both questioned forensic and disputed historical documents.

Osborn, Albert S. *Questioned Documents*, 2nd ed. Montclair, N.J.: Patterson Smith, 1978. Classic text recommended as much for its treatment of the basics of document examination as for its wealth of technical information.

Rendell, Kenneth W. *Forging History: The Detection of Fake Letters and Documents.* Norman: Univ. of Oklahoma Press, 1994. Illustrates how to detect forged historical writings; illustrated with many facsimile signatures of historical personages.

Sillitoe, Linda, and Allen Roberts. *Salamander: The Story of the Mormon Forgery Murders.* Salt Lake City: Signature Books, 1988. A thorough discussion of the Mark Hofmann forgeries and murders with a section on forensic analysis of the forgeries.

8

SEROLOGY

Blood is the bodily substance most commonly found at the scene of a crime or on a person, clothing, or a weapon potentially associated with a crime. It is a highly complex substance containing many cells, proteins, enzymes, and inorganic materials. This discussion is divided into three sections: the *principles* of forensic serology, the *conventional analyses* of blood and other body fluids, and *DNA testing*. The case study is the O.J. Simpson case.

PRINCIPLES

The science that studies the properties and effects of serums—for example, the analysis of blood traces—is known as *serology*.[1] As a practical matter, the serologist studies the various types of biological evidence: notably blood, semen, saliva, perspiration, and fecal matter. (However, tests conducted on such bodily substances in order to detect foreign materials such as alcohol in blood or drugs or poisons in urine are usually carried out by other criminalists who are specialists in chemistry or toxicology. Their work will be discussed in the following chapter.)[2]

Tests to identify blood have been known since at least 1875. At about that time, microscopic examination had limited value because it could not be applied to *dried* blood, nor could it distinguish between human blood and that of other mammals. Neither could another test, the hematin test, in which blood is mixed with salt crystals and concentrated acetic acid and heated to form the characteristic rhombic crystals of the blood constituent hemin. The most important test at that time was the guaiacum test. The suspected blood was placed in water to which was then added guaiacum (a tree resin) followed by hydrogen peroxide; if blood was present, a blue color was formed. Although this test could not identify the kind of blood, whether human or animal, it could be applied to a very tiny amount of dried blood.[3]

In 1887, the first Sherlock Holmes story, A *Study in Scarlet*, had the fictional detective inventing a new blood test. Its approach was like that of the guaiacum test: to a tiny amount of suspected blood in water were added certain chemicals ("a few white crystals, and then . . . some drops of a transparent fluid"). A reaction was obtained if the substance was indeed blood ("the contents assumed a dull mahogany colour, and a brownish dust was precipitated to the bottom of the glass jar"). Holmes claimed that his new test was an improvement over the "clumsy and uncertain" guaiacum test and highly sensitive.[4] While the test is, of course, fictitious, it is quite likely that the Conan Doyle story "gave impetus to the development of improved methods in blood identification."[5]

A breakthrough in identifying human blood came as a development of the experiments of German physician Paul Ehrlich (1854-1915), which led to the establishment of a new scientific discipline, immunology. In 1891, Ehrlich demonstrated that an "immune" serum against certain toxic substances could be obtained by injecting the substance into an animal in small doses. An antiserum (a serum with antibodies) could subsequently be taken from the animal. (If the toxin and antiserum were together injected into another, healthy animal, the toxin would produce no ill effects.) Subsequently, the forensic use of precipitating antisera to help identify the species that produced a given blood specimen was discovered, almost simultaneously, by Schutze, August von Wassermann, and Paul Uhlenhuth.[6] The latter's work was particularly significant:

> Uhlenhuth recognized, as did others, the necessity of testing bloodstains on a wide variety of substrata and of assessing the results from bloodstains of different ages. For example, he studied the results from stains of different ages on wood, sand, and cotton, among other substrata. Uhlenhuth was aware of the importance of adequate controls in his analyses and kept petri dishes containing dried flakes of blood of known age to assist him in this respect. . . . Uhlenhuth was involved in many cases, from murder to rape to chicken stealing. He also engaged in what must be some of the earliest examples of quality assurance in

forensic science. He examined bloodstains sent to him by other professors of legal medicine, the species of which were known only to them, and obtained the correct result.[7]

Another breakthrough came at the beginning of the twentieth century when Karl Landsteiner demonstrated that the serum of certain persons would agglutinate (clump) the red blood cells of certain other persons because the red blood cells contain antigens or blood-group factors, and the blood serum contains antibodies. Two such antigens, A and B, were identified, along with two antibodies, anti-A and anti-B. If a person has one of the two antigens, his blood type is A or B, respectively; if both, the blood type is AB; if neither, it is O. (It follows that if one has the A antigen in his red blood cells, his serum cannot contain an anti-A antibody, which would clump his own cells. Thus, one with the A antigen has the anti-B antibody, and one with the B antigen has the anti-A antibody; also, a person having both A and B antigens has neither anti-A nor anti-B antibodies, and one having neither antigen A nor B has both the anti-A and anti-B antibodies in his or her serum.[8] Landsteiner and his colleagues also discovered additional blood groups: the MN system in 1927 and, in 1940, the Rh blood groups.[9] Over time, still other factors have been discovered that help individualize blood.

In 1915, Dr. Leon Lattes, a lecturer and assistant researcher at the Institute of Forensic Medicine at the University of Turin, Italy, developed a procedure to apply AB0 testing to bloodstains on cloth and other materials. To do this, he first carefully restored the dried blood and serum as closely as possible to its original liquid state—a process that involved weighing the stain by cutting it from the fabric and then subtracting the weight of a similar swatch. He then used the appropriate amount of a saline solution to dissolve the dried blood.[10] (A couple of Lattes's first cases were briefly mentioned in the introduction.) In 1932, Lattes reported on a test he had developed for identifying the antibodies present in dried blood flakes.[11]

Much earlier, in 1910, scientists had discovered that blood groups were among the traits that are inherited following genetic laws, and therefore they could be utilized as evidence in paternity suits as well as in other criminal investigations.[12] The most recent—and perhaps the greatest forensic breakthrough in the realm of forensic serology—occurred in 1985 when Alec Jeffreys and his colleagues at Leicester University discovered that portions of the DNA structure of certain genes are unique to each individual. Consequently, Jeffreys termed the process used to isolate and read these genetic markers "DNA fingerprinting."[13]

It is now known that the antigens of the ABO system are not limited entirely to the red blood cells. In fact, some 80 percent or more people have their blood-type antigens in most of their body fluids (including saliva and perspiration). Such individuals are known as *secretors*.[14]

In the following sections we discuss the identification and individualization of bloodstains and other serological evidence. The collecting and preserving of such evidence is treated in chapter 2, as is blood-pattern analysis.

CONVENTIONAL ANALYSES

Since Landsteiner's discovery of the ABO system, the scientific knowledge that represents the field of forensic serology has mushroomed. Excluding hormones and some other factors, now more than 160 antigens are recognized, as well as 150 serum proteins and 250 cellular enzymes. Among the antigens are erythrocyte antigens, which are very common and are classified as factors of the various primary blood group systems; less common erythrocyte antigens are classified as factors of secondary blood groups (see table 8.1).[15]

Although some of the old tests for blood have been greatly improved (such as microscopic methods and the hematin test) and some sophisticated new methods have been developed (including spectrophotometry and electrophoresis),[16] the standard procedures still largely apply. These include preliminary tests to identify blood (or semen etc.), followed by tests for species, blood grouping, and other factors. The goal is to reach— or at least approach as closely as possible—the individualization of a sample of blood or other biological evidence.

Preliminary Tests. For practical reasons, including speed and convenience, it is customary to first identify a suspect substance as blood before proceeding to more complex tests. Such preliminary tests are not conclusive. Most tests are for an enzyme in blood called peroxidase, and unfortunately there are substances that can give similar "false positive" reactions (such as horseradish, which may be found in such blood-resembling substances as shrimp sauce). For this reason, preliminary blood tests are called "presumptive" tests; that is, they are used for screening. If the test is positive, other tests are required to identify the substance as blood.[17]

TABLE 8.1.
Erythrocyte Blood Grouping Systems

PRIMARY		SECONDARY	
ABO	Auberger	En	Ot
MNSs	August	Gerbich	Raddon
Rh	Batly	Griffith	Radin
Lewis (Le)	Becker	Good	Rm
Lutheran (Lu)	Biles	Heibel	Stobo
P	Bishop	Ho	Swann
Kell	Bg^{a}Bg^{b}Bg^c	Ht^a	Torkilden
I	Box	Jn^a	Traversu
Duffy	Cavaliere	Kamhuber	Vel
Kidd	Chido	Lan	Ven
Diego	Chr^a	Levay	Webb
Dombrock	Cost	Ls^a	Wright
Xg	Dp	Marriot	Wolfshag
Yt	El	Orris	

Formerly, the benzidine test was the standard preliminary indicator of blood.[18] However, benzidine is highly carcinogenic, and another indicator, orthotolidine, is only somewhat less so.[19] Therefore, phenolphthalein is becoming increasingly used as an indicator of blood in what is known as the Kastle-Meyer color test. Phenolphthalein reagent is mixed with hydrogen peroxide and the suspected bloodstain; a positive reaction is the development of a deep pink color.[20] Like benzidine, this test is quite sensitive, and it is easily used as a test.

Another presumptive test that is commonly used in the field is the luminol test. As mentioned briefly in chapter 6 regarding a case involving a latent shoe print, invisible traces of blood may be developed with a luminol spray. The result is a very faint glow that requires almost complete darkness for observation. Ironically, the test is most effective on older stains. Not only may faint blood-produced shoe prints be discovered with luminol, but even the smear marks caused by wiping or mopping, in attempts to remove bloodstains, can often be "visualized" by use of this test. (As a cautionary note, some experience is required in interpreting the results of the luminol test. A pinpoint glowing that is sometimes observed is not always caused by blood; actual bloodstains generally produce areas of luminescence.)[21]

Once a stain is preliminarily identified as blood, it must be confirmed as such and determined to be of either human or animal origin. Both requirements may be satisfied by the precipitin test. It is not only specific

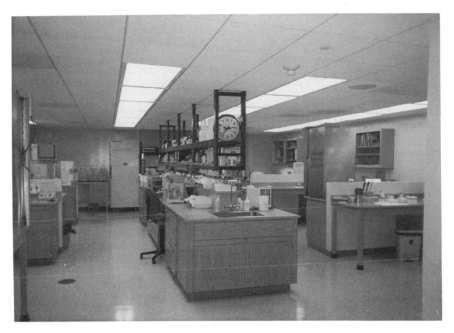

FIGURE 8.1. Typical forensic serology section of a modern crime laboratory. (Courtesy of Chris Watts, Orlando, Florida.)

for blood but also identifies the proteins in blood. The test is based on the fact that animals (typically rabbits) produce antibodies as a result of being injected with human blood. The resulting human anti-serum (or anti-human serum) reacts positively to human blood by forming a white cloudy precipitate. Appropriate controls (known standards of human blood and animal blood) are tested at the same time as the suspected blood in order to ensure that the antiserum is both accurate and specific.[22] Similarly, antisera may be prepared for other animal species, and the crime laboratory maintains antisera for domestic and other common species of animals such as dog, cat, horse, cow, and deer.[23]

Once the criminalist has determined that the blood is human, he then classifies it as to blood grouping. As discussed earlier, this is based on antigen-antibody reactions. Anti-A and anti-B sera are used to determine—by clumping of the red cells—whether the blood is type A, B, AB, or O. The most common groups are A and O, which constitute about 39 percent and 43 percent, respectively, of the general population. Therefore, it is quite likely that a significant number of individuals will have the same ABO blood type.[24]

On the other hand, the numerous additional blood factors discovered since Landsteiner permit a greatly increased characterization of blood. These include the Rh system, the MN system, and many others, including the following:

Adenosine deaminase	(ADA)
Adenylate kinase	(AK)
Erythrocyte acid phosphatase	(EAP)
Esterase D	(ESD)
Glucose-6-phosphate dehydrogenase	(G6PD)
Glyoxylase I	(GLO I)
Group-specific component	(Gc)
Haptoglobin	(Hp)
Hemoglobin	(Hb)
Peptidase A	(Pep A)
Phosphoglucomutase	(PGM)
6-Phosphogluconate dehydrogenase	(6PGD)
Transferrin	(Tf)

Some of the other constituents of blood—both plasma proteins and many enzymes—may be identified by electrophoresis. According to Inbau, Moenssens, and Vitullo in *Scientific Police Investigation:*

> In electrophoresis, the blood is placed in a well in a layer of agar on a glass plate. Voltage is introduced, causing the protein molecules that carry different electrical charges to separate from the blood and migrate across the agar. Suitable dyes are then applied to the gel after migration has been completed and the resulting pattern of separated proteins serves to identify the proteins that were present in the blood. Other proteins may require an antiserum that is specific for particular proteins, in which case the same procedure is followed but a trough is cut into the gel parallel to the direction of migration and the specific antiserum is added in the trough after the separation of the blood proteins. The antiserum diffuses from the trough and reacts with the serum proteins that have separated along a curved line in the gel. The resulting pattern is then interpreted in order to identify the proteins present. The use of antiserums to identify serum proteins that have been separated by voltage in a gel is known as immuno-electrophoresis. Enzymes present in blood may also be separated by electrophoresis and identified; the process involves a chemical reaction induced by the enzyme.[25]

Using a number of these factors together greatly narrows the frequency that a given blood sample can be expected to occur in a population.[26] The frequency in a given case can be determined by multiplying the percentages of occurrence for each. For instance, if a bloodstain is determined to be type A, which is known to appear in about 42 percent of the population, and is then determined to contain PGM 1 (one of the three common types of an enzyme known as phosphoglucomutase), which has an appearance rate of about 58 percent, then the origin of the stain can be reduced to approximately 24 percent of the public (42% x 58% = 24.36%).[27]

Like bloodstains, seminal stains often must be identified by the forensic serologist, usually in rape cases. Semen may be identified microscopically by the presence of spermatozoa, elongated microscopic structures having a rounded head and a long flagellate tail. Unfortunately, spermatozoa are very brittle when dry and disintegrate easily with handling or washing. The preferred method of simultaneously locating and identifying a seminal stain is by means of the acid phosphatase color test (acid phosphatase is an enzyme secreted by the prostate gland). The test utilizes an acidic solution of sodium alpha-naphthylphosphate and a dye, Fast Blue B. First a piece of water-moistened filter paper is rubbed over the suspected stain, thus transferring acid phosphatase, if it is present, onto the paper. Then a drop or so of the test solution is added. The appearance of a purple color indicates the presence of the enzyme. A negative reaction is interpreted as meaning semen is absent.[28]

Although other substances—notably fungi, contraceptive creams, and vaginal secretions—also react positively to the acid phosphatase test, the difference is that usually none of the other substances reacts as fast as does seminal fluid. If the reaction time is under thirty seconds, the presence of semen is indicated. Other methods may be used to confirm the identification. (It should be kept in mind when microscopically searching for spermatozoa that some individuals have a very low sperm count while others may have none as a result of a vasectomy.) One method of confirming the identification depends on the person being a secretor—one of the approximately 80 percent of the population whose blood type appears in semen and other body fluids. Typing the preliminary identified stain as A, B, AB, or O serves the dual purpose of confirming the presence of semen and providing further information on the person to whom it belonged. Today, the usual means of confirming the presence of semen when spermatozoa are not found is to test for a distinctive protein known as p30.[29]

Other biological evidence in addition to blood and semen includes saliva, perspiration, vaginal secretions, urine, feces, and vomitus. Except for vaginal secretions, these types of evidence are rarely encountered.

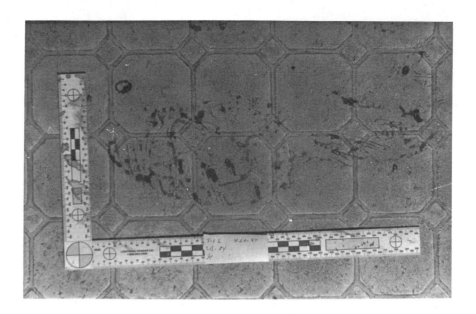

FIGURES 8.2 AND 8.3. Bloodstain evidence such as the blood shoe impression shown here (fig. 8.2) is examined by the forensic serologist in an effort to determine whose blood made the impression. The serologist removes samples for serological testing and may then treat the stained impression with amido blue-black dye to enhance it. The resulting enhanced impression (fig. 8.3) is photographed and then examined by a shoe impression analyst.

According to one forensic text, their main value lies in the possibility that they were produced by a secretor and that that person's blood type might be determined. (This might be possible even with a saliva stain on a cigarette or on the flap of an envelope.) Vomitus may yield evidence about a person's last meal and can sometimes provide an indication of the person's medical condition.[30] (A case involving tests of feces is briefly discussed in chapter 3.)

Another type of conventional serological analysis, although not usually relevant to a criminal investigation, is the determination of paternity. This may be accomplished by means of the blood-group systems of the suspected parents and the offspring. However, routine paternity testing utilizes the human leukocyte antigen (HLA) test rather than the ABO system. HLA relies on the identification of a complex system of antigens present on white blood cells. If a suspect cannot be ruled out as the father of a child on the basis of an HLA test, the odds are in excess of 90 percent that he is indeed the father. This test, combined with the ABO and another method (known as haptoglobin blood typing), can raise the odds to in excess of 95 percent. Currently, DNA testing, which is discussed in the following section, can increase the probability of paternity to more than 99 percent.[31]

DNA TESTING

The 1911 discovery of deoxyribonucleic acid (DNA)—the carrier of genetic information—caused scarcely a ripple in medical jurisprudence. But when Alec Jeffreys and his Leicester University colleagues discovered that portions of the DNA structure of certain genes are unique to each individual, they revolutionized forensic serology. Jeffreys coined the term "DNA fingerprinting" for their process of isolating and reading the DNA markers. That was in 1985. The following year, the first use of DNA technology in a criminal case occurred when Jeffreys was asked by police in his homeland to assist with a case. He was to verify the confession of a suspect that he had committed two rape-murders in the English Midlands.

Although the suspect's confession had the ring of truth, he confessed only to the *rape* of the second fifteen-year-old, whereas semen tests provided evidence of a dual killer. Also, the suspect, a kitchen porter, was a somewhat retarded seventeen-year-old whose blood tests were inconsistent with the semen tests. Jeffreys's new technology promised to resolve the matter. But whereas police thought the suspect would be linked to the two crimes, he was actually found to be innocent of both. According to one source, "On November 21, 1986, legal and forensic history was made when the teenage kitchen porter became the first accused murderer to be cleared as a result of DNA fingerprinting."[32]

The police were back to the proverbial square one. Then in early 1987 they made a decision to obtain a blood specimen from every male between the ages of sixteen and thirty-four who lived in a three-village area. Although the time required for the DNA process had been reduced from weeks to days, the process was still time-consuming. After two months and many hundreds of tests, the police got a break. They learned of a barroom conversation in which a man named Ian Kelly had said that he had been bullied into taking the test for another man. The latter, a man named Colin Pitchfork, claimed he had a record for indecent exposure that would cause the police to treat him badly. So, using fake identification, Kelly had given a blood specimen in Colin Pitchfork's name. Arresting both men, police then rushed a sample of Pitchfork's blood to Jeffreys's lab. This was the 4,583rd sample to be tested, and it was the last. On January 22, 1986, Colin Pitchfork pled guilty to both of the rape-murders and was sentenced to life imprisonment.[33] (In 1987, in another case in England, one Robert Melias drew the dubious distinction of being the first individual convicted of a crime—rape—based on DNA evidence.[34]) The advantages of DNA testing are even greater than the Colin Pitchfork case demonstrates. According to one discussion of the technology (as it pertains to the FBI laboratory):

> Like fingerprints, DNA can be used to make a positive identification. When an arm and hand were found in a shark's belly, the lab was able to identify the victim from a fingerprint. But fingerprints were of absolutely no value when part of a leg was found floating in a Florida river a few days after a woman college student had disappeared. However, because DNA from any part of the body is identical to DNA from any other part of the body, the lab was able to determine that this was not the missing student's leg. Even if a DNA specimen from the student hadn't been available, the examination could have been conducted by comparing DNA taken from the leg to DNA taken from a blood relative. Unlike prints, DNA is inherited, so a small segment of the DNA chain will be identical in all blood relations.[35]

Moreover, "One of the many advantages offered by DNA is that it's found everywhere white blood cells are found. Obviously, it's found in blood; it's found in semen in rape cases; it's in the skin found beneath a victim's fingernails after a struggle; it's in hair root follicles; it can even be found in saliva cells left on the mouthpiece of a telephone after a conversation."[36]

Basically, the DNA molecule looks like "a twisted ladder," or double helix, with "rungs" consisting of four sub-units called bases: guanine (G), adenine (A), thymine (T), and cytosine (C). The bases pair in certain predictable combinations—A with T and G with C—known as base pairs. While human DNA has in excess of three billion base pairs, only a small number determine those unique traits that attract forensic interest.[37]

The strands of DNA are folded into microscopic structures known as chromosomes and exist in all nuclei-containing cells such as those present in blood—except red blood cells, which lack nuclei. In addition to body fluids, DNA is present in tissues, hair roots, bone marrow, and tooth pulp. For forensic purposes, there are two basic DNA procedures in use: restriction fragment length polymorphism (RFLP) and polymerase chain reaction (PCR). (Another, nonnuclear, type of DNA—mitochondrial DNA—is discussed in chapter 11.)

Of the two procedures, RFLP is used more often in forensic work. Its disadvantages are that it requires a larger sample than PCR, is time-consuming and labor intensive, and utilizes radioactive reagents that require special lab procedures. Its great advantage is that it can individualize a specimen to a narrow portion of the population—possibly one person in billions.[38]

In contrast, PCR is faster and simpler to use and may be applied to exceedingly tiny samples—as small as a billionth of a gram of DNA. The technique is based on the way DNA strands reproduce themselves within a cell. With PCR, an enzyme known as DNA polymerase can be directed to duplicate a strand of DNA several million times. Unfortunately, the results of PCR are less dramatic than those of RFLP, being discriminatory on the order of one individual in thousands rather than the potential billions with RFLP.[39]

An important investigative capability utilizing DNA evidence is now under development by the FBI laboratory. This is a national computerized network that compiles the databases of various states that require sex criminals and sometimes other violent offenders to provide blood for DNA testing and recording. This computerized system—called the Combined DNA Index System (CODIS)—is already paying dividends. For example, CODIS linked a 1991 Miami rape case, in which police had not identified a suspect, with a man who had been convicted of sexual assault in Orlando in 1993.[40]

Courts have sometimes challenged DNA evidence on specific grounds— such as a private laboratory's failure to comply with suitable standards and appropriate controls. However, properly applied DNA testing is widely

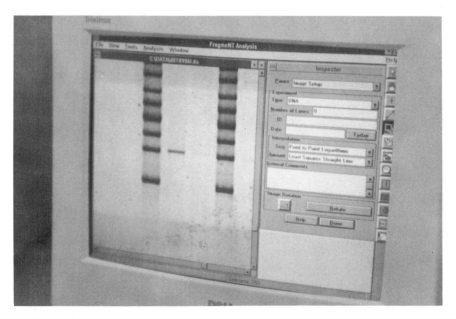

FIGURES 8.4 AND 8.5. A computerized system for viewing a DNA sample (fig. 8.4). The scanner attached to the system allows the forensic expert to scan the DNA x-ray film and visualize the results on a computer monitor (fig. 8.5). (Courtesy of Dr. Joyce Lee, Mesa Police Department, Mesa, Arizona.)

accepted as admissible under the standards set by *Frye* or *Daubert* (criteria established by the U.S. Court of Appeals and the U.S. Supreme Court, respectively, discussed in chapter 1). In its 1996 report on DNA evidence, the National Research Council wrote: "The state of the profiling technology and the methods for estimating frequencies and related statistics have progressed to the point where the admissibility of properly collected and analyzed DNA data should not be in doubt."[41] Attorney General Janet Reno commented: "Our system of criminal justice is best described as a search for the truth. Increasingly, the forensic use of DNA technology is an important ally in that search."[42]

CASE STUDY: THE O.J. SIMPSON TRIAL

Said to replace the Lindbergh kidnapping as "the trial of the century,"[43] the O.J. Simpson case has certainly been one of the most observed, studied, and debated events in history. From the discovery of two murders that have been described as butchery to the arrest of a high-profile "sports hero" as the suspected killer and his apparent attempt to flee the country, from the televised trial that became popularly termed a "circus" to the verdict that shocked a nation and contributed to its polarization along racial lines, the O.J. Simpson case threw into the spotlight of controversy the entire field of criminalistics, particularly the disciplines of crime-scene investigation and forensic serology. A subsequent civil trial seemed to many to redeem the perception of justice that was thwarted by the criminal trial.

Late in the evening of June 12, 1994, a resident of Brentwood, an affluent Los Angeles neighborhood, heard the agitated barking of a dog coming from a condominium across the alley. The condominium, on Bundy Drive, was the residence of Nicole Brown Simpson, the divorced white wife of black celebrity O.J. Simpson—a former professional football player, an occasional movie star, and a pitchman for Hertz Rental Cars. The dog was still barking when the neighbor finally went to sleep about eleven o'clock.

Later the bloodstained dog, Nicole's white Akita, was found wandering in the neighborhood. It ended up in the custody of a man named Sukru Boztepe, who had taken it from another neighbor. The dog led Boztepe to the body of its once-beautiful mistress, lying in a pool of blood outside her condominium. Boztepe asked his wife to phone 911, a call recorded at 12:10 A.M. on the morning of the thirteenth.

Arriving police secured the scene, posted guards around the perimeter, and strung yellow crime-scene tape across the various entrances to the grounds. Discovering two small, sleeping children in an upstairs bedroom, police carried them out a rear entrance so that they would not see the gruesome murder scene. They were taken to a police station and soon were picked up by an older half-sister, Arnelle (O.J. Simpson's daughter from his previous marriage).[44]

Lying near Nicole's body was that of a handsome twenty-five-year-old man named Ronald Goldman. He was a waiter from the Mezzaluna Trattoria, the restaurant where Nicole had dined earlier in the evening with her mother. Mrs. Brown had left her prescription eyeglasses at the Italian restaurant, and Goldman had delivered them to the care of Nicole, whom he knew. Investigators would conclude that Goldman's death was the result of being proverbially at the wrong place at the wrong time.

Police detectives went to Simpson's Rockingham Avenue estate only a few miles away to inform him of his ex-wife's death. They could rouse no one, although there were lights on inside the walled premises, and they discovered blood on the door of Simpson's white Ford Bronco, which was parked outside. Fearing that there could be more victims inside, the police climbed over the wall to investigate. (A municipal judge would later uphold their right to do so without a search warrant.)

Inside the grounds, a trail of blood led to the front door of the house. In an apartment in the guest wing at the rear of the estate, Detective Mark Fuhrman encountered Simpson's perpetual houseguest, mop-haired Brian "Kato" Kaelin. Kato told Fuhrman that he had heard thumping sounds outside his room earlier. The detective investigated and spotted a bloody dark-brown leather glove between the outside wall and a fence. The right-hand glove matched a left-hand one found at the scene of the crime at Bundy Drive. (It would eventually be learned that in December 1990 Nicole Simpson bought two pairs of such Aris Lights gloves—albeit of unspecified color or size—at Bloomingdale's in Manhattan.)

Simpson was not at the estate when detectives arrived. He had earlier been taken by limousine to the Los Angeles airport and from there had flown to Chicago on business. Reached at his hotel and informed of the death of his ex-wife, Simpson agreed to return to Los Angeles on the first possible flight. About noon on the day following the slayings, LAPD detective Philip Vannatter noticed a bandage on the middle finger of Simpson's left hand. Vannatter had a vial of Simpson's blood collected by a police nurse.[45]

FIGURE 8.6. Author Joe Nickell at the scene, on Bundy Drive in Los Angeles, of the murders of Nicole Brown Simpson and Ronald Goldman. (Courtesy of Fritz Stevens, Los Angeles, California.)

Simpson's blood was at issue because the killer had apparently received a wound during the savage murders, somewhere on his left hand. This was indicated by blood at the Bundy crime scene.[46] To the left of bloody shoe prints were blood drops that trailed alongside. These began on the walkway near the murder victims and led away from the condo, continuing beyond the rear gate into a driveway. Association with the departing footprints showed that the blood drops—four located on the walk and a fifth discovered in the driveway—were shed by the murderer. Later analyses revealed that the blood contained Simpson's DNA. (Since this blood evidence was collected before a sample of Simpson's own blood was obtained, the latter could not have been used to plant such evidence *in situ* as would later be suggested by Simpson's defenders, although it would have been possible later to switch the collected evidence for faked evidence.)[47] Simpson gave differing accounts of how he had cut himself.

Further evidence against Simpson was found and collected the following day, June 14, from his Bronco. This included bloodstains, subsequently shown to have Simpson's DNA, located on the driver's side carpet, the steering wheel, the instrument panel, the center console, and the driver's side wall. In a partial bloody shoe imprint on the driver's side was found Nicole's DNA, and on the center console was found a stain containing both O.J. Simpson's and Ron Goldman's DNA. (Additional bloodstains that contained a mixture of the DNA of Simpson, his ex-wife, and Goldman were collected later on August 26.)

The trail of Simpson's blood continued onto his estate. Drops were collected from the driveway and from the foyer and master bedroom of his home on June 13. A pair of dark socks was also recovered at this time, but bloodstains on the socks were not noticed until later, giving conspiracy theorists the opportunity to suggest that the blood was planted. Also collected from the estate was the bloody glove discovered by Detective Fuhrman.[48]

While the forensic serologists at the California Department of Justice and private contractor Cellmark of Maryland were conducting the sophisticated DNA tests on the various bloodstains, LAPD detectives found other evidence that pointed to O.J. Simpson. At the Bundy scene, for example, a blue knit watch cap was found that yielded thirty-four hairs, some bleached blonde like those of Nicole and others that were Negroid.

Then there was Simpson's record of abusing his former wife—an alleged forty-three incidents over the seventeen years of their turbulent relationship. Tapes and transcripts of various 911 calls, including a recorded incident on New Year's Day 1989 that showed a frightened Nicole in fear for her safety. When on that occasion police arrived about 3:00 A.M., Nicole was hiding in the bushes outside her house, clad in only a pair of sweat pants and a bra. When she refused to continue with charges against her husband, the city attorney filed misdemeanor charges of spousal battery against him. Also, when Nicole's safety deposit box was opened after her death, investigators found evidence documenting her abuse at the hands of Simpson. This included color photographs recording her bruised face, diaries filled with entries not only on beatings but also on being stalked by Simpson, and letters from him apologizing for his behavior and stating that he wanted her back. The box also included her new will, made only one month before her murder.[49]

Other evidence against Simpson concerned his lack of alibi during the "time window"—the period during which the murders could have been committed. Limo driver Allan Park told investigators that he was unable to reach Simpson on first arriving at the estate to pick him up for the trip to the airport. He rang the buzzer repeatedly. Then, while waiting outside, he saw a black man who fit Simpson's description come into view inside the estate and walk into the home. He saw lights come on, and, when Park again tried the intercom, Simpson answered the buzzer saying, "Sorry, I overslept, and I just got out of the shower. I'll be down in a minute."[50]

On June 17—just five days following the slayings and the day after he attended Nicole's funeral, accompanied by their children and other family members—O.J. Simpson was charged with murder. He was notified that a warrant for his arrest had been issued. While his attorney Robert Kardashian was on a telephone upstairs in Simpson's home, Simpson and his longtime friend A.C. Cowlings fled. They led police on a sixty-mile, slow-speed "chase" along area freeways. Simpson left behind a note addressed "To whom it may concern" and hinting at ending his life. He stated in part: "Don't feel sorry for me. I've had a great life, great friends. Please think of the real O.J. and not this lost person." He concluded: "Peace and love. O.J."[51]

Simpson and Cowlings fled not in Simpson's white Bronco (which had been impounded as evidence) but in Cowlings's identical 1994 vehicle. Cowlings drove during the "chase" while Simpson lay in the back seat with a pistol held to his head. At one point he used his cellular phone to dial 911 and—threatening to kill himself—asked to talk to his mother. Shortly after 8:00 P.M., Simpson exited the Bronco at Rockingham and surrendered to police. After being searched, he was allowed to talk by phone with his mother. He was also permitted to use the bathroom and to drink some orange juice before he was taken downtown for booking. He was assigned booking number BK4013970. When Cowlings's Bronco was searched, police discovered a loaded gun, cash totaling thousands of dollars, a passport, and a fake beard. Although Simpson would claim they were going to the cemetery where Nicole had been buried, in fact they had passed the exit that would have led them there and instead were heading south on the San Diego Freeway toward Mexico. The money, passport, and other items indicated to many that Simpson had intended to flee the country. This apparently aborted flight was widely interpreted as virtual proof of guilt.[52]

On June 20, however, Simpson entered a plea of not guilty to double charges of murder in the first degree. Four days later, a grand jury investigation into the Simpson case was halted by the judge, Cecil Mills, who blamed extensive media coverage for exposing the panel to "potentially prejudicial matters not officially presented to them by a district attorney."[53]

Without a grand jury indictment, a preliminary hearing was held before Judge Kathleen Kennedy-Powell. This hearing began on June 30. On July 18, Judge Kennedy-Powell ordered O.J. Simpson to stand trial on the two first-degree murder charges. On July 20, Simpson announced through his business manager-attorney that he was offering a five-hundred-thousand-dollar reward to anyone who could provide information as to the "real killer or killers"—an action that provoked bitter laughter in some circles. When asked to enter a plea two days later at his arraignment, Simpson responded, "Absolutely, 100 percent not guilty."[54] At this time, Superior Court Judge Lance A. Ito was named to preside over the trial. By then, most of Simpson's "Dream Team" of defense attorneys were in place. It was first headed by Robert Shapiro, but Simpson later gave the lead position to African American attorney Johnnie Cochran. They were joined by famed defense attorney F. Lee Bailey, Harvard University Law professor Alan M. Dershowitz, and lawyer and DNA expert Barry Scheck, among others.

On July 27, Sharon Rufo, the mother of Ronald Goldman, filed a civil lawsuit against Simpson alleging wrongful death in the killing of her son. This civil suit would be expanded and actively pursued following the criminal trial. There were few other public developments until September 2, when the State of California filed a motion requesting that the jury be sequestered so that they would not be influenced by extensive media coverage. On the ninth, the district attorney's office announced that the state would not seek the death penalty in the case but would instead ask for Simpson to be sentenced to life without parole. (Apparently the prosecutors felt that Simpson's tremendous popularity would preclude any jury convicting him if that meant a sentence of death.) On September 26, jury selection was begun in the case.

By November 3, the twelve-member jury had been selected and was sworn in. During the following month, twelve alternate jurors were also chosen and were seated on December 8. In a move calculated to trim up to a year off the proceedings, during which time Simpson would have remained in jail, the defense dropped its legal challenge to the admissibility of the DNA evidence. This occurred on January 4, 1995, and one week later the jury was sequestered in preparation for the trial to begin. On January 24, the trial began with an opening statement by Deputy District Attorney Marcia Clark, and the following day Johnnie Cochran gave the first opening statement on behalf of the defense. On the last day of January, the first witness—911 operator and dispatcher Sharyn Gilbert—took the stand, part of the prosecution's opening salvo of evidence about Simpson's violence and threats against his wife.

The prosecutors—primarily Marcia Clark and Christopher Darden—strategically avoided mentioning DNA during their opening statements because the evidence was considered so utterly convincing and overwhelming that it would have the greatest impact later in the trial. The term "DNA Wars" came to be used to describe the critical battle between prosecution and defense experts over the reliability of the DNA evidence, including such issues as evidence collecting, laboratory handling, and interpretation of results. In an unprecedented development for criminal trials, three different labs performed the DNA analyses in the case. All three labs (Cellmark, California Department of Justice, and the LAPD) independently determined that DNA in the blood drops beside the bloody footprints matched O.J. Simpson's DNA. The Restriction Fragment Length Polymorphism (RFLP) analysis yielded a 1 in 170 million match of the questioned blood to that of Simpson, and the Polymerase Chain Reaction (PCR) test gave a 1 in 240,000 match to Simpson on four of the five drops, with a 1 in 5,200 match on the fifth.[55]

Although, as we have seen, this blood could not have been planted at the crime scene, the defense countered with the testimony of Dr. Henry Lee, a renowned criminalist. Lee observed that there were blood smudges on the packaging of one blood drop. He testified that this indicated the blood was not dry when it should have been, and hinted darkly that there was "something wrong." Lee discounted several innocent possibilities for the smudging. However, according to one knowledgeable commentator:

> There were significant problems with this argument that swatches had been switched, even beyond the audacity of the conduct alleged. The DNA found in the blood drops was heavily degraded, and there was no reason why swatches dabbed in O.J. Simpson's pristine blood sample with a preservative in it would have been degraded. And why would rogue police officers, who according to the defense's own theory truly believed that Simpson was guilty, have removed evidence that they would have thought would point directly to him? Besides, switching swatches would have required not simply substituting five planted swatches for the five blood drops collected, but many more than five swatches, since some of the blood drops were sufficiently large that they were collected not just on one but on several swatches, later tested by DNA analysis. The jump from the statement that there was "something wrong" to the conclusion that multiple swatches had been switched between June 13 and June 14 was a great leap with little basis in logic, reason, or evidence.[56]

The defense also tried to discredit the evidence of the five drops of blood by advancing a scenario that the blood—having been stored in plastic bags in a hot lab truck for several hours—had become severely "degraded." But, according to Harlan Levy, a prosecutorial expert on DNA:

This scenario required a highly unlikely series of events. Degradation would have had to have been so severe that the DNA disappeared entirely in each of five separate blood drops. Next, a transfer would have had to have taken place in which the blood present in Simpson's blood sample made its way into each of the blood drops themselves. Finally, there would have had to have been a failure in five separate controls, each designed to determine whether contamination had taken place. In each instance, for each blood drop, contamination would have had to have jumped onto the blood drop being tested, but declined to jump onto a control that would have revealed its presence. In an alternative explanation for this third eventuality, Barry Scheck argued to the jury that there was no reason to assume that the controls were handled properly in a laboratory that was a "cesspool of contamination," although [criminalist] Yamauchi testified in detail to his proper use of the controls.

Levy continues:

No one of these three events—total degradation, a transfer of DNA, and failure of five controls—was a strong possibility. The defense had built factual support for its own case by bringing out statements from the prosecution's own witnesses establishing that each of these occurrences was a theoretical possibility. It was unlikely that any one of these possibilities had occurred; there was less than a chance that all three had coincided. Besides, traditional blood tests that are not susceptible to contamination also typed to Simpson on one of the blood drops, reflecting a 1 in 200 person match to the former football player, indicating that contamination had not in fact taken place. This was powerful proof that there had been no contamination, but it could easily have been missed or discounted by a jury overwhelmed and troubled by extensive and complex scientific testimony about contamination.[57]

Nevertheless, the defense continued to challenge the other blood evidence, arguing "contamination" whenever possible, and otherwise hinting at—or directly suggesting—outright conspiracy and fraud to frame an innocent man. The prosecution was further inhibited by the technological complexity of DNA testing. According to one source: "Prosecutors struggled to make the mystifying material sound less like scientific gobbledygook and more like carefully examined evidence and meticulously assembled information that pointed a damning finger of guilt at the defendant. There were times nevertheless when jurors appeared to be too preoccupied with other matters or about to doze off while lawyers and witnesses droned on for weeks about such mind-boggling matters as Hardy-Weinberg equilibrium theory, population restructuring, P.C.R., R.F.L.P., D1S80, and D.Q.-alpha."[58]

The prosecution rested its case on July 6. Four days later, Arnelle Simpson, Simpson's daughter from his first marriage, was the leadoff witness for the defense. Among other things, she testified that her father's rheumatoid arthritis was so bad that he had given up tennis, and presumably by implication could not have committed the murders. This is an absurd argument on its face, all the more so when considering that, not long before the murders, Simpson actively participated in a fitness video.

The defense soon played what has become known as "the race card"—defense attempts to take advantage of racial animosities, particularly against the LAPD, and to appeal to the largely black jury by suggesting that racist cops had conspired to frame Simpson. Defense attorneys were able to take advantage of Mark Fuhrman's denial of having used the "N-word," proving with a tape recording that he had indeed used such a racially offensive epithet. By extension, the defense intimated, Fuhrman might also have lied about other things. Specifically, they suggested, he might have planted evidence—such as the incriminating bloody glove found at the Rockingham estate. This led some court watchers to remark that the proceedings had become the "Fuhrman trial"—an attempt to shift the trial's focus from the question of Simpson's guilt to the character and racial attitudes of Detective Fuhrman.[59] The defense never produced evidence that Fuhrman had, in fact, faked anything whatsoever in the Simpson case. Through all of this, interest in the proceedings mushroomed to the extent that around the country even gas stations installed mini television sets so that their customers could keep up with courtroom activities while they were filling their gas tanks.[60]

In the end, the "Dream Team" did not put Simpson on the stand. When asked by the court to confirm his decision not to testify, he included in his response a surprise statement of his innocence, outraging the prosecution as well the Goldman family. But the jury had nevertheless heard the statement. In retrospect, it seems to have been just what the predominantly black jury wanted to hear. Astonishingly, the jury deliberated less than four hours to reach a verdict. Since the evidence against Simpson was so profuse and so profound, many assumed that the hasty verdict was guilty. But when at 10:00 A.M. on October 3, 1995, the verdicts of "not guilty" on both counts were read, America was outraged—at least, white America was. Black America largely cheered the acquittal of one of its heroes, but, even there, opinion has since shifted toward guilt as more and more evidence has been brought forward: two groups of bloodstains—one on a gate at the murder scene and another on a sock from Simpson's bedroom—were particularly controversial. Those on the gate were not collected until July 3, and while the socks had been taken into evidence on June 13, the bloodstains had gone unnoticed until August 4. These late discoveries raised suspicions in the minds of Simpson's defenders. Additionally, a defense expert testified that the blood in both instances—that on the gate and that on the socks—contained a chemical called EDTA, which is used to preserve blood. This was cited as proof that the questioned bloodstains came from a collected specimen, since EDTA is not a constituent of blood. The prosecution, of course, having bloodstains in profusion, had no need to fake these stains. Their experts countered that laundry detergents, soft drinks, or other substances could cause positive results in EDTA testing, and an FBI expert testified that there was, in fact, no actual preservative in the blood on either the rear gate or on the socks.[61]

Other controversial evidence focused on shoe prints at the crime scene. The FBI's expert, William Bodziak, identified the bloody trail of prints as having been made by designer Italian shoes of Bruno Magli brand, size twelve. Such shoes are expensive and rare. (Simpson denied owning such shoes, but subsequently numerous photographs were discovered and presented at the civil trial showing O.J. Simpson wearing the distinctive Bruno Magli shoes with their particular tread pattern.) The other controversial footprints, identified by criminalist Henry Lee, were imaginary, according to Bodziak. One was actually an impression left by tools when the concrete was originally poured for the walkway. Bodziak also determined that an imprint on the envelope that contained Mrs. Brown's glasses was not that of a shoe, based on a computer check of the thousands of shoe impressions in the FBI's data bank.[62]

There was much incriminating evidence that was not presented at the criminal trial. For example, there was the money, passport, disguise, etc. found in A.C. Cowlings's Bronco after the infamous low-speed "chase." In addition, Judge Ito ruled the following evidence admissible, yet the jury heard none of it:

1. A 1982 incident in which O.J. allegedly smashed photos of Nicole's family, threw her against a wall, and threw her and her clothes out of the house.

2. A 1985 call to a private security officer in which a crying, puffy-faced Nicole alleged that O.J. had bashed her car with a baseball bat.

3. A 1987 incident in Victoria Beach in which O.J. allegedly struck Nicole and threw her to the ground.

4. A 1989 incident in which O.J. allegedly slapped Nicole and pushed her from a slow-moving car.

5. Several statements made by O.J. in 1993 and 1994. These include one incident in which O.J. showed a friend the secret back way into his ex-wife's home. "Sometimes she doesn't even know I'm here," Simpson allegedly said.

6. Several allegations of "stalking behavior," including two times that O.J. allegedly followed Nicole to restaurants when she was with another man, as well as one incident in which he claimed to have observed them having sex on a couch in her home.

7. A 1994 incident in which O.J. saw Nicole with Ronald Goldman and another man having coffee and allegedly stopped his car and angrily motioned her to come over. Judge Ito said that the episode was relevant because it connected Simpson to Goldman and was "evidence of jealousy and motive."[63]

More and more evidence against Simpson came out at the civil trial in which the Goldmans' and the Browns' suits were consolidated into a single case. The plaintiffs were in a better position than the prosecutors had been in the criminal trial in at least two respects: first, the requirement was only for proof by a preponderance of the evidence rather than proof beyond a reasonable doubt, and second, the jury's verdict did not need to be unanimous, but only nine of the twelve had to agree in order to resolve the matter. In fact, on February 4, 1997, a unanimous jury found O.J. Simpson responsible for the deaths of Nicole Brown Simpson and Ronald Goldman. Although appeals were expected, a measure of justice had finally arrived.

NOTES

1. Vernon J. Geberth, *Practical Homicide Investigation*, 2nd ed. (Boca Raton, Fla.: CRC, 1993), 584.

2. Fred E. Inbau, Andre A. Moessens, and Louis R. Vitullo, *Scientific Police Investigation* (New York: Chilton, 1972), 114.

3. Samuel M. Gerber, "A Study in Scarlet: Blood Identification in 1875," in *Chemistry and Crime: From Sherlock Holmes to Today's Courtroom*, ed. Samuel M. Gerber (Washington, D.C.: American Chemical Society, 1983), 31-35.

4. Arthur Conan Doyle, *The Complete Sherlock Holmes* (Garden City, N.Y.: Garden City Books, 1930), 6-7.

5. Gerber, "A Study," 35.

6. P.H. Whitehead, "A Historical Review of the Characterization of Blood and Secretion Stains in the Forensic Science Laboratory, Part One: Bloodstains," *Forensic Science Review* 5.1 (June 1993): 39.

7. Ibid.

8. Inbau, Moenssens, and Vitullo, *Scientific Police Investigation*, 119-20.

9. Gerber, "A Study," 34.

10. Jurgen Thorwald, *Crime and Science: The New Frontier in Criminology* (New York: Harcourt, Brace & World, 1966), 38-45.

11. Whitehead, "A Historical Review," 40.

12. Gerber, "A Study," 34.

13. Richard Saferstein, *Criminalistics: An Introduction to Forensic Science*, 5th ed. (Englewood Cliffs, N.J.: Prentice Hall, 1995), 383.

14. Ibid., 360-61.

15. Henry C. Lee, "Identification and Grouping of Bloodstains," in *Forensic Science Handbook*, Richard Saferstein, ed. (Englewood Cliffs, N.J.: Prentice-Hall, 1982), 268; Lawrence Koblinsky, "Bloodstain Analysis: Serological and Electrophoretic Techniques," in Gerber, *Chemistry and Crime*, 90. (The table is taken from the same source.)

16. Ibid., 271-83.

17. Inbau, Moenssens, and Vitullo, *Scientific Police Investigation*, 116-17.

18. Ibid.

19. A.J. Fisher, *Techniques of Crime Scene Investigation*, 5th ed. (New York; Elsevier, 1992), 224.

20. Saferstein, *Criminalistics*, 354.

21. Fisher, *Techniques of Crime Scene Investigation*, 224-25.

22. Inbau, Moenssens, and Vitullo, *Scientific Police Investigation*, 117; Saferstein, *Criminalistics*, 355.

23. Inbau, Moenssens, and Vitullo, *Scientific Police Investigation*, 117-19.

24. Ibid., 119-20; Fisher, *Techniques of Crime Scene Investigation*, 234-35.

25. Inbau, Moenssens, and Vitullo, *Scientific Police Investigation*, 121.

26. Lee, "Identification," 297-324; Saferstein, *Criminalistics*, 363; Michael Kurland, *How to Solve a Murder: The Forensic Handbook* (New York: Macmillan, 1995), 143.

27. Saferstein, *Criminalistics*, 362-63.

28. Ibid., 373.

29. Ibid., 373-75.

30. Charles R. Swanson Jr., Neil C. Chamelin, and Leonard Territo, *Criminal Investigation*, 4th ed. (New York: McGraw-Hill, 1988), 98.

31. Saferstein, *Criminalistics*, 372.

32. Colin Evans, *The Casebook of Forensic Detection* (New York: John Wiley, 1996), 60-61.

33. Ibid., 61-63.

34. Edward Connors et al., *Convicted by Juries, Exonerated by Science: Case Studies in the Use of DNA Evidence to Establish Innocence after Trial* (Washington, D.C.: U.S. Dept. of Justice, 1996), 4.

35. David Fisher, *Hard Evidence* (New York: Dell, 1995), 184.

36. Ibid., 184-85.

37. Fisher, *Techniques of Crime Scene Investigation*, 231-32.

38. Ibid., 232.

39. Ibid.; Saferstein, *Criminalistics*, 398-99.

40. Saferstein, *Criminalistics*, 405-6.

41. Connors et al., *Convicted by Juries*, 6.

42. Janet Reno, in ibid., iii.

43. Clifford Linedecker, *O.J. A to Z: The Complete Handbook of the Trial of the Century* (New York: St. Martin's/Griffin, 1995).

44. Ibid., passim. Except as otherwise noted, information for this case study is taken from this encyclopedia-format book.

45. Ibid.

46. Harlan Levy, "O.J. v. DNA," in *Postmortem: The O.J. Simpson Case*, Jeffrey Abramson, ed. (New York: Basic Books, 1996), 105-16.

47. Ibid., 106.

48. Ibid., 110-15.

49. Linedecker, *O.J. A to Z*, passim.

50. Ibid., 19. See also 171, 231.

51. Ibid., 224.

52. Ibid., 30, 36-37, 90, 216-17.

53. Ibid., 154.

54. Ibid., 18.

55. Levy, "O.J. v. DNA," 106.

56. Ibid., 107.

57. Ibid., 108-9.

58. Linedecker, *O.J. A to Z*, 75.

59. Ibid., 93, 160, 185.

60. Ibid., 97.

61. Levy, "O.J. v. DNA," 109-10, 112-14; Linedecker, *O.J. A to Z*, 80.

62. Linedecker, ibid., 32-33, 39.

63. Elizabeth M. Schneider, "What Happened to Public Education about Domestic Violence?" in Abramson, *Postmortem*, 79-80.

RECOMMENDED READING

Connors, Edward, et al. *Convicted by Juries, Exonerated by Science: Case Studies in the Use of DNA Evidence to Establish Innocence after Trial.* Washington, D.C.: U.S. Department of Justice, 1996. Includes commentaries by Attorney General Janet Reno, Barry Scheck, and others.

Fuhrman, Mark. *Murder in Brentwood.* New York: Zebra, 1997. The inside story of the O.J. Simpson case by the first investigator on the scene of the bloody murders.

Lee, Henry C. "Identification and Grouping of Bloodstains." In *Forensic Science Handbook*, Richard Saferstein, ed. Englewood Cliffs, N.J.: Prentice-Hall, 1982. Thorough discussion of forensic analyses to determine blood groups.

Levy, Harlan. "O.J. v. DNA: What the DNA Really Showed." In *Postmorten: The O.J. Simpson Case*, Jeffrey Abramson, ed. New York: Basic Books, 1996. Substantiates the prosecution's view of the DNA evidence, disparaging conspiracy theories.

Linedecker, Clifford. *O.J. A to Z: The Complete Handbook of the Trial of the Century. New York*: St. Martin's/Griffin, 1995. Reference guide to most aspects of the O.J. Simpson murder trial.

Noble, Deborah. "Forensic PCR: Primed, Amplified, and Ready for Court." *Analytical Chemistry* (Oct. 1, 1995). Excellent discussion of PCR which can be used to test tiny amounts or degraded samples of DNA.

Whitehead, P.H. "A Historical Review of the Characterization of Blood and Secretion Stains in the Forensic Science Laboratory, Part One: Bloodstains." *Forensic Science Review* 5.1 (June 1993). Identifies significant achievements in bloodstain testing over the past century and a half.

9 CHEMISTRY

INTRODUCTION

When the public, along with Dr. Watson, were introduced to Mr. Sherlock Holmes in 1887, they found him a student in a hospital laboratory, busily developing a chemical test for bloodstains. When it was proposed that he and Watson share lodgings (Holmes had his eye on "a suite in Baker Street"), the great sleuth confessed: "I generally have chemicals about, and occasionally do experiments. Would that annoy you?" As it turned out, Holmes would sometimes spend his day "at the chemical laboratory" (when he was not "in the dissecting rooms" or out taking long walks "into the lowest portions of the city"). As well, "his hands were invariably blotted with ink and stained with chemicals." Although Watson, in assessing Holmes's knowledge of various subjects assigned him a "nil" in both literature and philosophy, he ranked his knowledge of chemistry as "profound." Even a description of the "untidy" Baker Street quarters reflected the importance of chemistry to "the scientific detective" (as Holmes styled himself): There were "the scientific charts up on the wall, the acid-charred bench of chemicals, the violin-case leaning in the corner." In fact, as Watson mildly complained: "Our chambers were always full of chemicals and of criminal relics which had a way of wandering into unlikely positions, and of turning up in the butter-dish or in even less desirable places."[1]

The emphasis on chemistry is appropriate. Chemistry is one of the five major classifications of science (along with biology, physics, earth science, and space science). Chemistry is the science that treats the *composition* of substances and their *transformation*, the reactions by which they undergo change into other substances. Its analytical aspect deals with determination of the exact composition of substances. Called *chemical analysis*, this aspect has two major divisions: *qualitative analysis*, which seeks to determine *what* elements or compounds may be present, and *quantitative analysis*, which attempts to determine relative *amounts* of each.[2]

A century before Sherlock Holmes, the end of the eighteenth century saw the origins of modern chemistry. By 1814, with the publication of his treatise on poisons, Mathieu Orfila established toxicology—a branch of forensic chemistry—as a legitimate scientific discipline. And decades before Holmes's appearance, medico-legal experts recognized the need for the chemical identification of substances such as bloodstains.[3] As one of the major subdivisions of science, it is not surprising that chemistry continues to be a mainstay of criminalistic procedures. Not only can chemistry have application as a major forensic discipline in its own right, but it often has value in areas that generally are not associated with the flask or retort. For example, the fingerprint expert may use a chemical such as ninhydrin to develop a latent thumb impression; the firearms examiner may employ chemical methods for detecting lead residue around a bullet hole or apply an etching reagent to restore a serial number; the specialist in questioned documents may utilize thin-layer chromatography to identify a certain brand of ink; the microscopist may use microchemical techniques in the analysis of trace evidence; and so on, as we have seen in the preceding chapters. In this chapter, we look at forensic applications of chemistry in investigations regarding *drugs, toxicology, arson,* and *explosives.* The case study is the World Trade Center bombing.

DRUGS

The problems associated with drugs are complex and are best left to clinical, sociological, and criminological discussions. Here our primary concern is that of the forensic chemist; namely, with identification of illegal drugs. Such drugs may be divided into four major types: narcotics, depressants, stimulants, and hallucinogens.[4]

Narcotics. From the Greek narkotikos ("to benumb"), the word *narcotic* has come to be applied loosely to a variety of drugs, including a stimulant, cocaine, and a mild hallucinogen, marijuana. True narcotics are drugs that relieve pain through depression of the central nervous system and sleep. Most narcotics are derived from opium, which in turn comes from the oriental poppy plant (*papaver somniferum*). The plant—which grows primarily in Asia, Mexico, and the Balkans—yields the opium in the form of a milky substance that is obtained by slitting open the unripe poppy pods. The raw opium is then pressed into cakes, which are naturally dark brown or black. This material is used to make the opium that is sold on the market, a dark-brown extract produced by a process of boiling, fermenting, and roasting. The finished product may be chewed, eaten, or smoked. (Medicinal forms include an opium solution known as *laudanum,* which was a popular analgesic during the eighteenth and nineteenth centuries, and liquid, powder, and granulated forms of modern medicinal opium.)[5]

Other narcotics include such opium derivatives as morphine, codeine, and heroin. *Morphine* (commonly as morphine sulphate) is an important medical analgesic (or painkiller) that accounts for about 12 percent of raw opium. The legal form is typically a one-grain white tablet; the illegal drug is usually in the form of a white powder. It may be taken orally, a method that addicts regard as wasteful, or "mainlined" (injected directly into the bloodstream). This is accomplished by dissolving the drug in water placed in a spoon and heated; the solution is commonly injected by a needle attached to an eyedropper with a rubber band (instead of the conventional syringe).

Although *codeine* is actually found in opium, it typically is made synthetically from morphine. Its strength is only one-sixth that of morphine, so it is not popular as a street drug. Its primary use is as a cough suppressant in cough syrup. In this form it is sometimes a source of abuse by juveniles, although such abuse tends to be sporadic and minimal—of little significance to the overall drug abuse problem.

Heroin is a derivative of morphine (made by reacting it with acetyl chloride or acetic anhydride). By far, it is the drug that most commonly results in cases of narcotic addiction. It is usually a white, crystalline powder, but may occasionally be found in tablet or cube form. It is used in essentially the same way as morphine—a user's "kit" consisting of a spoon (often with a bent handle), a syringe or eyedropper with affixed needle, a constriction band, and a book of matches, possibly accompanied by a candle.[6]

Among the narcotic drugs that are not derived from opium—drugs known as synthetic opiates—the most common is methadone (methadone hydrochloride). Uniquely, although it is related to heroin, methadone tends to eliminate an addict's craving for heroin while at the same time causing only minimal side effects. It is sometimes used in treating heroin addiction to relieve withdrawal pains, even though methadone

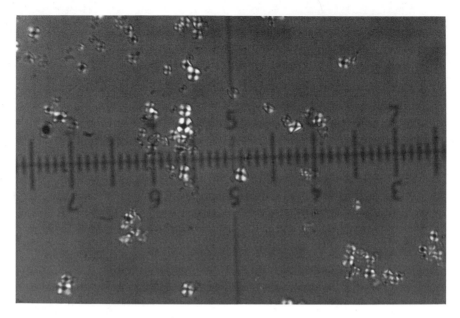

FIGURE 9.1. Photomicrograph of starch grains viewed through a polarizing light microscope (100x). Starch is often used to dilute or "cut" such drugs as heroin and cocaine.

itself is capable of being addictive. (Methadone hydrochloride is known by various trade names including Adanon, Amidon, Amidone, Dolophine, and Methadon.) Other synthetic opiates include meperidine hydrochloride (Isonipecaine, Demerol, and others), and dihydromorphinone hydrochloride (Dilaudid).

Depressants. Drugs that have a depressing action on the central nervous system are called depressants. The most widely used such drug—indeed, "the most widely used and abused drug in the world today"[7]—is alcohol. An estimated 50 percent of those committing fatal assault are intoxicated, as are those committing suicide. Some 40 percent of the victims of homicidal assault are likewise in a state of intoxication.[8] (Alcohol will be discussed further in the next section).

A major category of depressant drugs is represented by *barbiturates*, commonly called "downers" because of their ability to cause one to relax, experience a feeling of well-being, and lapse into sleep. A proper dose may be medically beneficial whereas an overdose may produce death. In fact, barbiturates are often the drug of choice for those intending suicide. Barbiturates are also addictive and produce a withdrawal syndrome more severe than any other drug. Barbiturates are derivatives of barbituric acid (first synthesized over a century ago by German chemist Adolf von Bayer), and include phenobarbital, sodium Amytal, Nembutal, and Seconal.

In addition to alcohol and barbiturates, depressants include the non-barbiturate known as methaqualone (Sopor and Quaalude), a powerful sedative; tranquilizers such as chlorpromazine; so-called "mild tranquilizers" including meprobamate (Miltown and Equanil), chlordiazepoxide (Librium), and diazepam (Valium); and chloral hydrate (a.k.a. "knockout drops"). There are also certain volatile solvents that can depress the central nervous system when "huffed" or sniffed—as in "glue sniffing"—including toluene, gasoline, naphtha, trichloroethylene, and methyl ethyl ketone.[9]

Stimulants. In contrast to depressants, stimulants are drugs such as *amphetamines*, commonly known as "uppers" or "speed," that stimulate the central nervous system and provide increased alertness followed by a decrease in appetite. Serious abuse often begins when either amphetamine, or its chemical derivative methamphetamine, is injected intravenously.

Another major stimulant is *cocaine*, a drug stimulant once used medicinally as a local painkiller, but promoted by Sigmund Freud in the 1880s as a source of "exhilaration and lasting euphoria."[10] An extract of the leaves of the coca plant (*Erythroxylon coca*), cocaine is a white crystalline alkaloid powder. Pre-Columbian Andean Indians chewed the coca leaves or made a tea to relieve pain or fatigue; today the processed cocaine is typically sniffed or injected. In extreme cases, side effects include hallucinations and paranoia.

Hallucinogens. Mood- and perception-altering drugs that may produce hallucinations are called hallucinogens. They include such natural drugs as *peyote* (the "buttons" of the peyote cactus), *mescaline* (an alkaloid derived from peyote), and *psilocybin* (an extract of Mexican mushrooms). Synthetic hallucinogens include LSD (d-lysergic acid diethylamide), DMT (dimethyltriptamine), and STP (an unidentified substance such as atropine). LSD was especially popular as a "mind-expanding" drug of the sixties, and it is reportedly "the most powerful hallucinogen yet developed."[11]

A relatively mild hallucinogen comes from the female hemp plant, which produces a resin that is sold in pure form and is known as hashish. The leaves (and often the seeds, stems, etc.) are the most common form and represent the drug *marijuana* (*cannabis sativa*). Whereas hashish is usually smoked in a special pipe, marijuana is typically made into cigarettes called "reefers" or "joints" and is "the most widely used of the illicit drugs."[12]

Drug tests. The forensic identification of "controlled substances" and other drugs begins with screening tests. Field tests are usually spot tests whereby a small amount of the questioned substance is placed in the well of a spot plate and one or more chemical reagents is added. A lack of reaction is interpreted as an absence of the drug being tested for; a positive reaction, usually a color change, is an indication that the substance *may* be the drug in question. Further tests are required. (See table 9.1.)[13]

Other preliminary tests are microscopic analysis (especially useful in the case of marijuana and in drugs with microcrystals of distinctive shape and color) and ultraviolet spectrophotometry (based on comparison of the UV spectra with compilations of such spectra).[14]

Following the screening tests, which reduce the possibilities as to what the questioned substance may be, come certain confirmatory tests. According to Siegel:

> Since it is not possible for drug chemists personally to test all the millions of known substances against an unknown in a particular case to make sure that the test is specific, a confirmatory test must be *theoretically* specific; that is, the analyst can predict that it would not respond in the same manner to other untested substances. Within the realm of controlled substance, a confirmatory test must also be specific in *practice*. A drug chemist must be reasonably scientifically certain that the test used for confirmation is able to distinguish among all controlled substances and their isomers, and so on.[15]

The most common confirmatory tests are infrared spectrophotometry and gas chromatography–mass spectrometry. Infrared spectrophotometry is based on the tendency of even very similar substances to exhibit different infrared (IR) spectra. The unknown spectrum is compared to a "library" of spectra from known compounds. Gas chromatography—mass spectrometry is a powerful technique resulting from the combination of two analytical methods in a single instrument. It first separates the various components of a possibly complex drug mixture, then unequivocally identifies each of them.

TOXICOLOGY

The study of poisons is known as *toxicology* (from the Latin word *toxicum*, "poison"). The creation of forensic toxicology by Mathieu Orfila (1787-1853) is discussed in chapter 1. Originally, the science consisted largely of the attempted detection and identification of poisons in cases of sus-

TABLE 9.1.
Spot Tests for Drugs

Test	Drug	Reaction
Marquis	heroin, morphine, amphetamines	purple, orange-brown
Dillie-Koppanyi	barbiturates	violet-blue
Duquenois-Levine	marijuana	purple
Van Urk	LSD	blue-purple
Scott	cocaine	three-step test: blue-pink-blue

pected homicide. Today, suicides and "accidental" suicides (in which persons take only a moderate overdose to attract attention but later die of complications such as liver failure) "make up the bulk of the toxicological work carried on in the West."[16] Actually, that is true only of "classic" toxicology. By extension, the modern toxicologist has been assigned the job of testing blood and urine specimens for alcohol—"the most common substance tested for in most police laboratories"[17]—in suspected cases of DUI (driving under the influence, often called "drunk driving").

By further extension, testing may also need to be done for drugs. Considering the range of work of the modern toxicologist, this discussion is divided into the following topics: *alcohol and drugs* and *poisons in death investigations.*

Alcohol and Drugs. To measure the level of blood alcohol, samples of blood, breath, or urine may be taken, depending on the jurisdiction. Blood—the preferred specimen for either alcohol or drug testing—must be taken in a medically approved manner. Alcohol must not be used to clean the syringe since it might affect the test results. Additionally, a non-alcoholic cleaner (such as aqueous zephiran) should be used on the area of skin from which the sample is to be drawn. About 10 to 20 ml of blood should be obtained, using a container supplied for the purpose and containing an appropriate anticoagulant and preservative.

Urine samples to be tested for alcohol or drugs are collected after the subject has first voided the bladder and waited about twenty minutes. The urine is then collected in a container with an appropriate preservative. A collected sample of about 25ml is recommended. With both blood and urine, an officer should observe the procedure, mark the evidence appropriately, and submit it to the laboratory to be tested.[18] Breath

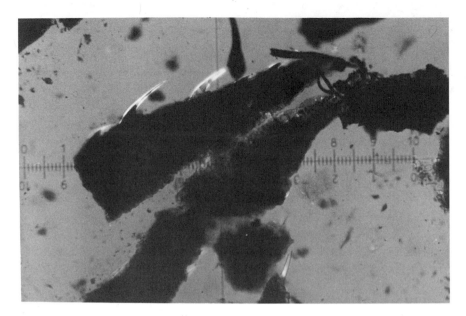

FIGURE 9.2. Photomicrograph of charred marijuana residue, specifically illustrating characteristic cystolithic hairs (100x).

samples, used for alcohol tests only, are not transported to the laboratory. Instead, instruments used for field collecting—such as the Breathalyzer, which is a sort of specialized spectrophotometer—also analyze the specimen.[19]

In general, the laboratory tests for blood can be applied to other biological samples, such as urine. The early methods of testing tended to be chemical, but today the most widely used analytical techniques for identifying alcohol in the forensic laboratory are gas chromatographic methods. These methods vary somewhat.[20] The basic procedure, like other uses of gas chromatography, involves vaporizing the specimen, injecting the resulting gases into an inert carrier—a gas such as nitrogen, which is swept through a special tube. The different constituents exit at different retention times that can be measured and are characteristic for known constituents. Analysis of blood or urine for alcohol requires only a single drop. In just minutes, an instrumental pen trace shows whether or not alcohol is present and, if so, in what amount.[21]

It should be noted that blood specimens obtained from a deceased person require special procedures, since bacterial action can produce ethyl alcohol during the process of decomposition.[22] In the case of biological fluids and tissues to be tested for drugs, the toxicologist takes

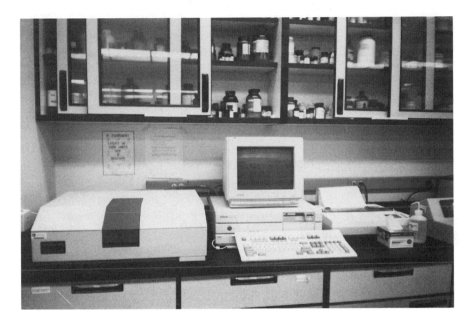

FIGURE 9.3. Ultraviolet and visible spectrophotometers such as this one are used to eliminate thousands of substances, leaving a narrow group which can be identified by such confirmatory tests as infrared spectrophotometry. (Courtesy of Chris Watts, Orlando, Florida.)

advantage of the fact that some drugs are acidic, like barbiturates and aspirin, while others are basic or alkaline, such as amphetamines, cocaine, methadone, and phencyclidine (PCP). Once the blood or other specimen has been extracted and separated into acidic and basic "fractions," there follows the two-step process of screening and confirmation. The common screening tests are the familiar thin-layer chromatography (TLC) and gas chromatography (GC), in addition to another screening tool called immunoassay, which is based on specific antigen-antibody reactions like those used in blood typing. (While antibodies that can react with specific drugs do not exist naturally, they can nevertheless be produced by mixing the given drug with a protein and then injecting the mixture into an animal, such as a rabbit, the blood of which will then contain antibodies for the given drug.[23]) As mentioned in the previous section on drugs, the preferred confirmatory choice is gas chromatography—mass spectrometry, described as a "one-step confirmation test of unequaled sensitivity and specificity."[24]

Poisons in Death Investigations. When a person dies under dubious or suspicious circumstances, or in any case where poison somehow comes

FIGURE 9.4. A gas chromatograph-mass spectrometer (located along the back wall) is used to separate and identify many substances encountered by the forensic chemist. (Courtesy of Chris Watts, Orlando, Florida.)

to be suspected as a possible cause of death, the pathologist sends samples of body tissues and fluids to the laboratory for analysis by a toxicologist. The samples should contain three ounces of blood, without preservative; all of the stomach contents (and in some cases, the stomach itself); one-half of the liver; both kidneys; all of the urine from the bladder; and one-half of the brain. The blood should be taken from both chambers of the heart as well as from other sites in the body. The lungs and entire brain are to be taken if a volatile poison is suspected (i.e., one that evaporates rapidly), and the spleen if cyanide is indicated. Even if the body is skeletonized, arsenic may be detected from the hair, nails, and bone (particularly a segment from the femur).(See Table 9.2.)[25]

Circumstances that may point to poison include corrosion or burns about the mouth, the presence of drugs or drug paraphernalia at the scene, or the fact that a victim was a potential target by virtue of being a burden on others. Therefore, any medicines or other substances found near the victim, along with any cups, drinking glasses, or other containers should be collected at the scene as potential evidence.[26] According to an experienced homicide investigator: "The most important fact to keep in mind is that the scene examination and investigation into the events

TABLE 9.2.
Major Types of Poisons

---

1. IRRITANTS
    1.1. mineral acids and alkalis
        1. sulfuric acid (vitriol)
        2. hydrochloric acid (muriatic acid)
        3. nitric acid (aqua fortis)
        4. ammonia
    1.2. organic acids
        1. carbolic acid (phenol, creosote, Lysol)
        2. hydrocyanic acid (prussic acid)
        3. oxalic acid (an industrial bleach)

2. METALLIC POISONS
    2.1. arsenic (e.g., insecticides, rat poison)
    2.2. antimony
    2.3. mercury
    2.4. lead (e.g., from old paint)

3. ORGANIC OR VEGETABLE POISONS
    3.1. alkaloids
        1. opium derivates (e.g., morphine)
        2. belladonna group (atropine etc.)
        3. strychnine (used in medicines such as cathartics)
    3.2. nonalkaloids
        1. barbiturates
        2. chloral hydrate ("Mickey Finn")

4. GASES
    4.1. hydrogen sulphide (toxic agent in sewer gas)
    4.2. phosgene (formed through decomposition of chloroform)
    4.3. carbon monoxide (as from auto exhaust)
    4.4. carbon dioxide (can cause asphyxiation in places where heating appliances
        are used without proper ventilation)

5. FOOD POISONING
    5.1. bacterial
        1. salmonella group (e.g., from insufficiently cooked poultry products)
        2. B. botulinus (botulism, e.g., from improperly canned foods)
    5.2. nonbacterial (such as certain kinds of mushrooms, like Amanita muscaria)

---

leading to the death must be thorough and complete. The medico-legal autopsy will determine the type and quantity of the poisonous substance involved. However, determination of the mode of death will be based on the police investigation at the scene."[27]

An early toxicologist visiting a modern laboratory would no doubt recognize such familiar items as test tubes and flasks, and might even identify a microscope as such, but would be bewildered by the array of strange machines that seem animated with lives of their own. As one source describes the array:

> Today's toxicologist is largely dependent on a battery of analytical techniques using various forms of chromatography—a method of separating, measuring and analysing chemical mixtures. The machines used are complex, but in basic terms chemicals are fed into an electronic machine which the toxicologist adjusts to his requirements; the amounts of each substance present emerge in the form of either a graph or, in ultra modern instruments, a computer print out.
>
> There are several methods for different jobs: gas chromatography, paper and thin layer chromatography, and most recently high pressure liquid chromatography. This can detect tiny traces of such formerly 'difficult' substances as LSD in the urine, and monitor the strength present by a process known as radioimmunoassay. Thus in a sample of blood, urine, or other liquid the measurement of drugs or poison present using these methods need be no more than .000000001 grams. No toxicology laboratory can now do an up-to-date and thorough job without at least two of these appliances.[28]

In addition to routine tests for drugs or alcohol and the analysis of tissue samples for suspected poisons such as arsenic or cyanide, the toxicologist may play an important role in the investigation of fires. For example, if the decedent were to have a less-than-fatal level of carbon monoxide in his or her blood, it could indicate that the person was dead before the fire began. (Elevated levels of carbon monoxide also can be caused by smoking and by living in a polluted environment.) Furthermore, any of various chemicals found in the decedent's blood could indicate that the person was breathing them in with the smoke from the fire and was therefore alive at the time, and the nature of the chemicals may indicate that an accelerant was used (for example, the presence of nitrous oxides would point to nitrogen-containing fuels).[29]

ARSON

Incendiary fires (those deliberately set) quite often involve the use of accelerants—something used to promote and spread the fire. An accelerant may be a solid, a liquid, or even a gas. *Solid accelerants* range from such simple materials as paper or trash to solid chemical incendiaries such as highway flares, black powder, or certain improvised mixtures such as paraffin and an oxidizer, or sugar and chlorate. *Liquid accelerants* are more frequently encountered and include such flammable liquids as petroleum products (gasoline, kerosene, fuel oil, etc.), alcohols, paint thinners, industrial solvents, and ether. Such liquids may be sloshed about an area or used in a firebomb such as a "Molotov cocktail." *Gas accelerants*, such as natural gas and propane, are less commonly used; however, the simple disconnecting of a gas line is one readily available source for this type of accelerant.[30]

An incendiary device is typically used with the accelerant. This device may be something as simple as a candle—anything that combines a means of ignition with a time delay. (The usual method is for the candle to be used to light a "fire set"—a cluster of paper or other material, possibly soaked in a low-volatility flammable liquid, such as oil, and arranged at the base of the candle so as to ignite when it burns down.) Other incendiary devices range from simple fuses to highway flares, smoke grenades, and "Molotov cocktails," from an open book of matches with a lit cigarette placed behind, to spontaneously combusting chemical mixes and even electronic devices, possibly incorporating such items as kitchen timers, wires, batteries, etc.[31]

The usual rules of documenting, collecting, and preserving evidence apply to arson investigations, and this added advice is offered by Charles R. Midkiff of the Bureau of Alcohol, Tobacco, and Firearms (ATF):

> As with other types of criminal investigations, the key to successful examination of arson evidence begins with the scene investigator. He must recognize, collect, and properly package the evidence if the laboratory is to be of assistance. When flammable liquids are involved, improper packaging can lead to loss subsequent to collection. Petroleum products such as gasoline are well known for their volatility. If permeable containers such as paper or plastic bags are used for packaging of the evidence, evaporation through the container walls can occur and the vapors will go undetected. Therefore, the trained arson investigator will package his evidence only in clean vapor-tight containers such as screw-cap glass jars, metal cans, or unused paint cans with tight fitting lids. With proper packaging, loss of the accelerant is avoided and potential cross-contamination of exhibits is eliminated. When the laboratory is provided with properly collected and packaged evidence, examinations can be made with reasonable chance of success.[32]

Some of the materials used to start a fire may be destroyed during the resulting conflagration. Wires, batteries, and other items may survive relatively intact, however, as may the neck of a bottle containing remains of a wick, and careful searching may even reveal a pool of wax remaining

FIGURE 9.5. Arson fires are quite often ignited by means of simple devices, such as the household timer shown here.

from a candle. Obviously, the place to look for traces of any incendiary device and/or fire set is at the *point of origin* of a fire when that can be determined by the arson investigator. Other telltale residues may be soap (used as a gelling agent to convert a flammable liquid into a napalm-like mixture), characteristic ash residues (such as the decomposition products of a flare's fuel and oxidizer), or residues of certain commercial materials that are sometimes employed as incendiaries (including thermite, welding compositions, and smokeless gunpowder).[33]

The laboratory analysis of evidence from suspected arson fires is largely accomplished by means of the gas chromatograph, considered "the most sensitive and reliable instrument for detecting and characterizing flammable residues."[34] The usual procedure is to heat the airtight evidence container (as indicated earlier, this may be an unused paint can) so that volatile residues will be driven off the collected material and be trapped in the container's "headspace" (enclosed air space). This vapor can be removed with a syringe and injected into the gas chromatograph. The analytical chromatogram that results may be compared to known standards to prove that, for instance, the presence of gasoline is indicated.[35] A cautionary note:

Although many laboratories rely on headspace examinations alone, this may lead to a number of "false negatives," or failure to detect a flammable liquid present in the evidence. This situation occurs most often with products of low volatility such as kerosene or fuel oil. There will often be too few peaks in the chromatogram for an identification to be made or even for a determination that an accelerant is present. The lower vapor pressure of these products does, however, limit loss through evaporation at the fire scene and if not completely burned, they will be recoverable from the physical evidence. Gasoline by contrast, on most nonporous surfaces is rapidly lost either through burning or evaporation during or subsequent to the fire. Absorbent surfaces retain even gasoline well.[36]

In the case of *unburned* residual liquids that have been collected, these may be recovered by separating them from the substrate. A variety of techniques is available for this, depending primarily on the type of substrate rather than the suspect liquid. The most widely used lab technique for this is steam distillation. Other techniques are vacuum distillation, solvent extraction, solvent rinsing, and air flushing. (Solvent extraction, for example, as its name implies, utilizes a suitable solvent such as carbon tetrachloride or chloroform to remove certain accelerants from porous materials. The solvent used for the extraction must be eliminated before testing. This is done by evaporation, and using a solvent with a low boiling point helps to minimize loss of the suspect flammable liquid.)[37] Since many of the solid incendiary materials are also used in explosives, the applicable tests will be described in the following section.

EXPLOSIVES

In the past, bombings tended to be rare and to be aimed at limited targets. In the "Roaring Twenties," dynamite bombs were largely the province of political anarchists who targeted the rich and powerful or of workers in the fierce battles to establish unions. In the 1930s, stench or stink bombs were used by gangsters to intimidate the owners of movie theaters and restaurants into selling out to the Mob; explosive bombs were used if the owners refused. Rarely were bombs aimed at the general public.[38]

That began to change on November 1, 1955, with the midair explosion of United Airlines flight 629, just eleven minutes into a departure from Denver's Stapleton Airport. The wreckage of the DC-6B was scattered over an area five miles in any direction. A characteristic odor persuaded investigators that a bomb was responsible. Proving it was another matter. The FBI and Civil Aeronautics Board (CAB) joined forces in an approach that would continue in its basic form into the present time. A surveyor was commissioned to establish a "base line" through the wreckage in the approximate line of flight. Using perpendicular lines, he bisected this line at thousand-foot intervals to create a giant grid for identifying the location of the various pieces of wreckage. The relatively undamaged tail section was found about one and a half miles from the nose section and engines. "It seemed to have been neatly sliced from the rest of the plane, as though by a knife," according to one account.[39] A warehouse at Stapleton was placed under guard, and the various pieces of the plane were taken there for reassembly. The pieces were carefully fastened with wire to a wooden mock-up of the plane. The result, except for the occasional bare patch, was a gaping hole in the tail section on the right side. The skin of the fuselage was bent outward around the hole, indicating that the explosion took place inside the plane at that location, corresponding approximately to the number 4 cargo pit. Several of the metal fragments bore gray and black deposits. Analysis at the FBI laboratory showed the deposits consisted of sodium carbonate, together with traces of sulfur and nitrate compounds—the expected residue from dynamite. A fragment of a six-volt battery was also identified. Investigation soon led to Jack Graham, a man who had taken out several large travel-insurance policies on his mother, who had perished in the crash. A search of his home turned up yellow-insulated wire like that collected at the crash site and identified as part of the detonator. Although he denied it, his wife had seen him place a "gift" in his mother's suitcase just before the flight. Eventually Graham confessed, saying, "The number of people killed made no difference to me. It could have been a thousand." Although he subsequently recanted, he was convicted on additional evidence and was executed in Colorado's gas chamber on January 11, 1957.[40]

Explosives may be divided into two types: high and low. *Low explosives* are those that yield an energy wave transmitted with a relatively low velocity, only a few thousand feet per second. They emit a low-frequency sound as well—typically described as a "pop," "puff," or "boom"—and cause relatively small damage. Examples of low explosives are gasoline and gunpowder. In contrast, *high explosives* have velocities up to twenty-five thousand feet per second, emit a piercing high-frequency blast, and create an explosive crater. Dynamite and nitroglycerin are examples of high explosives.[41]

Another way of classifying explosives is by source. *Military explosives* are usually based on TNT (trinitrotoluene) or an explosive called RDX. Another military explosive is known as PETN. *Commercial explosives* range from black powder and ammonium nitrate-fuel oil (ANFO)—both used for blasting—to various explosive gels and high explosives like dynamite and nitrostarch. *Improvised explosives* frequently include simple mixtures like homemade black powder and fuel mixed with an oxidizer. The improvised variety are usually low explosives, requiring confinement in order to have effective blasting power; rarely are high explosives improvised.[42]

Bomb-scene searches are similar to those of arson. Both involve a focal point (a point of origin for a fire, a "blast seat" for an explosion). Both involve evidence that is largely destroyed, and any remaining traces need to be searched for diligently. Such traces include explosive residues, metal fragments from a pipe (from a pipe bomb), remnants of a fuse or blasting cap, bits of wire and/or insulation, pieces of electrical tape, fragments of batteries, and various parts of clocks or timers.[43] According to the ATF's Charles R. Midkiff:

> Because the construction of a bomb is largely dictated by the available materials, the ingenuity and skill of the maker, the investigator cannot anticipate what may be recovered in a careful scene search. If, however, he does a thorough job of collecting and identifying the evidence, the laboratory will often be able to provide sufficient information to allow the reconstruction of a device virtually identical to the one placed by the bomber. Despite destruction and disruption, few other crime scenes provide as much potentially valuable physical evidence as is available at the bomb scene. This evidence, when subjected to a thorough laboratory examination accords information which can lead to the development or identification of a suspect, and the association of this suspect with the manufacture and placement of the device.[44]

The FBI's Explosives Unit maintains an impressive explosives reference collection, described as "perhaps the most complete collection in existence of bits and pieces that might be used in an explosive device." Included are actual samples of domestic and foreign batteries, detonators, various blasting accessories, road flares, timers, radio-control devices and components—the list goes on and on. Manuals and catalogues allow technicians to identify a particular type of pipe, for example, or other possible component of a bomb. The unit also has a computerized

FIGURE 9.6. A mobile robot, typical of those used by law enforcement agencies to recover or disarm explosive devices. (Courtesy of Orange County Sheriff's Department Hazardous Devices Unit, Orlando, Florida.)

bomb reference file, which enables experts to link components and techniques with cases having similar characteristics.[45]

When debris from the bomb scene arrives at the crime lab, the first step is a microscopic examination. This is performed primarily to locate particles of the explosive that have not been consumed. Black and smokeless powder, for example, are comparatively easy to identify by their characteristic colors and particle shapes. Next, acetone may be used to extract soluble explosives from the debris, and the extract is then concentrated and subjected to various spot tests and the versatile thin-layer chromatography. Nitroglycerin, RDX, and PETN are readily identified in this manner. Additionally, another chromatographic technique—high-performance liquid chromatography (HPLC)—may be employed to detect trace explosive evidence.[46]

Other chemical and instrumental tests may be performed, given the numerous potential constituents of explosive devices. One device that is sometimes used to screen evidence is known as an "explosives detector"—the most widely used such detector being the Vapor Trace Analyzer (VTA), a special-purpose gas chromatograph that offers both high sensitivity and high selectivity for particular explosive compounds.[47]

When sufficient quantities of explosives are recovered, the identifications may be confirmed either by infrared spectrophotometry (which produces a distinctive "fingerprint" pattern for organic substances) or

FIGURE 9.7. The robot's command vehicle, displaying the robot and other equipment relating to the emergency response to hazardous devices. This includes the equipment necessary to receive television pictures from the robot and guide it to the explosive device to disarm it. (Courtesy of Orange County Sheriff's Department Hazardous Devices Unit, Orlando, Florida.)

by x-ray diffraction (which gives a characteristic diffraction pattern for inorganic explosives.[48]

The following case study, the bombing of the World Trade Center, demonstrates some of the complexities and successes of a major modern explosives investigation.

CASE STUDY: THE WORLD TRADE CENTER BOMBING

Just eighty-six feet shorter than Chicago's Sears Tower, the Twin Towers of the World Trade Center—the second- and third-tallest buildings in the United States—loom far above the remainder of the seven-building complex that occupies sixteen acres in lower Manhattan. The towers stand a quarter-mile tall, are anchored in solid bedrock, and sway gently in the breeze—moving as much as four feet in either direction in high winds. One of the steel towers is crowned with an observatory, the other with a restaurant. Far below, the massive basement of the Center contains a five-level parking garage with space for up to two thousand vehicles.[49]

On Friday, February 26, 1993, at 12:17:37 P.M., a massive explosion occurred in the garage. The street shook, the towers began to sway, and

FIGURE 9.8. A valuable tool for the forensic chemist is the infrared (IR) spectrophotometer. Many substances can be identified by the IR technique since they have a unique IR spectrum. (Courtesy of Robert Miller, Phoenix Police Department, Phoenix, Arizona.)

smoke rolled up a ramp from the garage, pouring outside. Almost immediately, people in the towers above began piling into the stairwells seeking to escape the upstairs offices, which were beginning to fill with the smoke that was carried upward through the vents. Frantic calls went to 911, whose operator patched calls through to the fire department and police. One call was from a bleeding man, trapped in the garage. Many people were trapped in the Port Authority's underground train station and offices. Four Port Authority employees were found dead there; a fifth victim died of a heart attack following the blast. The last body was not recovered until March 15. Some fifty thousand people evacuated the building, most by way of smoke-filled stairwells. Of the more than one thousand who were injured, most suffered from smoke inhalation.[50]

As reports of the explosion went out from the news media, the cause of the blast was not immediately apparent. One source reported a fire; some authorities thought a transformer had exploded. Nevertheless, as a precaution, the FBI's Explosive Unit chief, Chris Ronay, dispatched special agent/bomb expert Dave Williams to New York.[51] He was at the scene the following morning but was scarcely prepared for what awaited him. As he recalled later:

> It was surreal. It was like walking into a bizarre cave. The explosion had ripped through five levels of the parking garage. Slabs of concrete as large as basketball courts, eighteen inches thick, were just hanging in midair, and they'd suddenly break loose and fall several stories and the whole building would begin to shake. Steel beams were broken and twisted. There had been sixteen hundred cars in the garage when the bomb went off, and their alarm systems had been activated. Wherever you looked headlights were flashing and lights were blinking and all you could hear were sirens and bells and whistles. Their electrical systems were sparking and starting pockets of fire, there was smoke everywhere, and some of the bravest people I've seen were risking their own lives to search through the rubble for victims. In two seconds one of the most modern buildings in the world had become a dark, dangerous cavern.
>
> The building was still moving. We put a spacer in a half-inch crack and walked a hundred feet, and by the time we got back the spacer was gone and the crack was six inches wide.[52]
>
> . . . .
>
> I knew we had a bomb the moment I walked in. When I came down the ramp, I saw big pieces of debris that had been thrown more than seven hundred feet from the seat of the blast, and there just wasn't enough natural gas or methane in there to generate that kind of force. In the unit we have a policy of not attributing an explosion to a bomb until we have physical evidence of that bomb, but . . . when I was asked about it by [New York State] Governor [Mario] Cuomo I told him, "If it looks like a duck and walks like a duck, it's a duck." We found out soon enough that this was the largest improvised explosive device—meaning somebody made it, as opposed to a military bomb—that had ever functioned in the United States.[53]

When the structural engineers had assessed the danger of the towers' possibility of collapse and had installed braces on the underground steel columns, the crime-scene technicians of the FBI and ATF donned hard hats, face masks, and rubber boots and began to climb through the dangerous, twisted wreckage. They began by searching for explosive residues. They even hung from cranes, swabbing the walls for possible traces. But no residues were found. Instead there was an unusually high percentage of urea and nitric acid—clues that helped lead to the eventual identity of the explosive, a chemical mixture called urea nitrate. An improvised or "homemade" explosive, it was known to have been used only once before, in 1988 when four college students were killed while making a pipe bomb.[54]

Dave Williams had already concluded that a large amount of explosive had been used, and that it had been brought in in a truck or van. Identification of the vehicle was the next logical step—one that had brought success in many "car-bomb" cases. By early Monday morning, two technicians, one from ATF and the other from the New York Police Department's (NYPD) bomb squad, discovered a large piece of a rear frame rail whose damage suggested that it had been part of the bomb—carrying vehicle or had been close to it when it went off. Fortunately, the piece bore a vehicle identification number (VIN). An NYPD detective used a toothbrush to clean away soot and then applied ordinary fingerprint powder to enhance the number. The number was immediately placed into the National Criminal Intelligence Center computer, and just as quickly the computer identified the vehicle. It was a Ford Econoline van, owned by the Ryder Rental Agency. Rented in New Jersey, it had been reported stolen by the man who had rented it, one Mohammed Salameh. His name was immediately recognized—he had been of interest to the FBI's New York office for some time.[55]

Because of the alleged theft, Salameh was attempting to reclaim his four-hundred-dollar deposit. The FBI hoped to follow him from the Ryder office when he returned for his deposit, thinking he might lead them to others, but, as so often happens, reporters learned of the situation and compromised that strategy by descending on the Ryder office. After this happened, agents abandoned their plan of attempting surveillance and arrested Salameh promptly after he left the Ryder office, where an FBI agent had posed as a manager. They then searched Salameh's residence, where they obtained links to other conspirators.[56] According to one source:

> Salameh's personal records, phone bills and things like that, led agents to Nidal Ayyad and Mahmud Abouhalima, who'd helped him buy the chemicals and make the bomb. When the details of the arrest were made public, the owner of a Jersey City storage facility contacted the Newark FBI office and said that the day before the bombing he'd seen four men loading a Ryder van. He'd checked their storage space and discovered a whole lot of chemicals. At the site investigators found about three hundred pounds of urea, 250 pounds of sulfuric acid, about a dozen bottles of nitric acid, two fifty-foot lengths of hobby fuse, a pump, a light blue Rubbermaid thirty-two gallon trash can, and six two-quart bottles of a substance that looked like onion soup, but turned out to be

homemade nitroglycerin. Sitting in that storage area was everything needed to make another three-hundred-plus-pound bomb.[57]

The hobby fuse told the experts that, instead of using an electric detonating system, the bombers had simply lit a fuse and left in a hurry. Confiscated pamphlets and a videotape described the precise method of making urea nitrate, and that function was connected to a fourth suspect, Ahmad Ajaj, a Palestinian with training in guerrilla warfare. Investigators even located the makeshift lab where the terrorists had mixed the explosive—a New Jersey apartment.[58] From various acid-burn marks and traces of nitroglycerin on the floor, Dave Williams concluded: "I'm firmly convinced that if these guys hadn't planted the bomb when they did, they would have ended up all over Jersey City in a lot of little pieces."[59]

Following the bombing, a letter had been sent to the *New York Times* claiming responsibility for the bombing on behalf of a group styling itself the Fifth Liberation Brigade. The FBI's DNA Unit was able to extract DNA from the saliva used to moisten the gummed flap of the envelope and matched it to Nidal Ayyad. Ayyad was a Palestinian-American and graduate of Rutgers University who was employed as a chemical engineer at a New Jersey chemical company.

As the bombing experts reconstructed the way the massive explosion was carried out, they came to the conclusion that the terrorists had mixed gallons of nitric acid with urea crystals in large metal drums, forming a gel-like mass that was then mixed with old newspapers and funneled into boxes. They had next produced homemade nitroglycerin. When the van was loaded, it contained a cargo of fifteen hundred pounds of urea nitrate. Stacked beside the boxes were three cylinders of compressed hydrogen, four containers of nitro, to each of which was attached a packet of gunpowder. Four twenty-foot fuses were connected to the gunpowder after being threaded inside surgical tubing to minimize smoke and slow the rate of burning. After twenty minutes, the lit fuses would have ignited the gunpowder, which would have detonated the nitro, which would have set off both the hydrogen and the urea nitrate paste.[60] The results were—tragically—already known.

Mahmud Abouhalima had fled the country but was captured by Egyptian authorities and turned over to U.S. officials. Another suspect, Ramzi Ahmed Yousef, remained at large. The four in custody went to trial, with opening statements beginning on October 4, 1993.[61] Among the problems the prosecution faced were those concerning evidence at the scene:

Even the chemical analysis had let them down. Bits of twisted metal were still being scraped and the shavings analyzed for chemical residue. So far, the results had been dispiriting. They were looking for a perfect crystal of nitroglycerin that could be chemically matched to what was left behind in the bomb factory. But the chemicals in the bomb had been almost completely consumed in the heat of the explosion. What remained were traces of nitrites, ammonium nitrate, and urea. Those were by-products of a urea-nitrate bomb, but they were also chemicals that saturated the remains of the parking garage. Ordinary car exhaust deposited nitrites. Fire extinguishers used to douse spot fires left ammonium nitrate residue. Before the explosion and during the rescue operation, maintenance workers had blanketed the area with a urea-based deicer. Because ruptured sewer pipes had soaked nearly every floor, urea from the wastewater was not hard to find.[62]

Nevertheless, the totality of evidence was convincing, and all four were convicted on a total of thirty-eight charges; each man was sentenced to 240 years in the penitentiary. The sentence was based on the judge's calculation of how many years the victims could have been expected to live.[63]

In the meantime, in June 1993, the FBI arrested a number of suspects in another terrorist bomb plot in New York. Many of them were connected to the mosque of Sheikh Omar Abdel Rahman, who was himself indicted on sedition charges by a federal grand jury on August 25. Secret audio tapes obtained by an FBI undercover agent, a former Egyptian army colonel named Emad Salem, showed the Sheikh advising the conspirators on various targets.[64]

NOTES

1. Arthur Conan Doyle, *The Complete Sherlock Holmes,* Garden City, N.Y.: Garden City Books, 1930), 6-11 ("A Study in Scarlet"), 94 ("The Sign of Four"), 444 ("The Musgrave Ritual"), and 1193 ("The Adventure of the Mazarin Stone").

2. *The Lincoln Library of Essential Information* (Buffalo, N.Y.: Frontier Press, 1946), s.v. "chemistry."

3. See chapter 1.

4. This discussion of drugs, except as otherwise noted, is based on Charles E. O'Hara, *Fundamentals of Criminal Investigation,* 3d ed. (Springfield, Ill.: Charles C. Thomas, 1973), 258-301; Richard Saferstein, *Criminalistics: An Introduction to Forensic Science,* 5th ed. (Englewood Cliffs, N.J.: Prentice Hall, 1995), 243-75; and Jay A. Siegel, "Forensic Identification of Controlled Substances," in *Forensic Sci-*

ence Handbook, vol. 2, Richard Saferstein, ed. (Englewood Cliffs, N.J.: Prentice Hall, 1988), 68-160.

5. O'Hara, *Fundamentals,* 262-63.

6. Ibid., 264-65.

7. Saferstein, *Criminalistics,* 256.

8. O'Hara, *Fundamentals,* 746.

9. See n. 4.

10. Sigmund Freud, quoted in Saferstein, *Criminalistics,* 260.

11. O'Hara, *Fundamentals,* 287.

12. Ibid., 275.

13. The table is compiled from information in Saferstein, *Criminalistics,* 269-70.

14. Siegel, "Forensic Identification of Controlled Substances," 82.

15. Ibid., 83-84.

16. Patrick Toseland, "Toxicology," in *Science against Crime,* ed. Yvonne Deutch (New York: Exeter Books, 1982), 72.

17. Barry A.J. Fisher, *Techniques of Crime Scene Investigation,* 5th ed. (New York; Elsevier, 1992), 343.

18. Ibid.

19. Saferstein, *Criminalistics,* 287-92; Yale H. Caplan, "The Determination of Alcohol in Blood and Breath," in Saferstein, *Forensic Science Handbook,* 623-41.

20. Caplan, "Determination of Alcohol," 608, 616.

21. Russell Stockdale, "The Forensic Science Laboratory," in Deutch, *Science against Crime,* 32-33.

22. Saferstein, *Criminalistics,* 296.

23. Ibid., 304, 352-53.

24. Ibid., 304.

25. Toseland, "Toxicology," 75; H. Ward Smith, *Laboratory Aids for the Investigator* (Toronto: Attorney-General's Laboratory, 1962), 21. (The table is an outline of the discussion in O'Hara, *Fundamentals,* 531-38.)

26. Toseland, "Toxicology," 71, 72; Vernon J. Geberth, *Practical Homicide Investigation,* 2nd ed. (Boca Raton, Fla.: CRC, 1993), 246-48.

27. Geberth, *Practical Homicide Investigation,* 248.

28. Toseland, "Toxicology," 75.

29. Charles R. Swanson Jr., Neil C. Chamelin, and Leonard Territo, *Criminal Investigation,* 4th ed. (New York: McGraw-Hill, 1988), 301.

30. Charles R. Midkiff, "Arson and Explosive Investigation," in Saferstein, *Forensic Science Handbook,* 222-24.

31. Ibid., 235.

32. Ibid., 224-25.

33. Ibid., 234-39. O'Hara, *Fundamentals,* 242.

34. Saferstein, *Criminalistics,* 327.

35. Ibid., Midkiff, "Arson and Explosive Investigation," 225-34.

36. Midkiff, "Arson and Explosive Investigation," 231.

37. Ibid., 231-32.

38. David Fisher, *Hard Evidence* (New York: Dell, 1995), 66.

39. Andrew Tully, *The FBI's Most Famous Cases* (New York: William Morrow & Co., 1965), 193.

40. Ibid., 191-201; Angus Hall, ed., *Crimes and Punishment*, vol. 7 (N.P.: Symphonette Press, 1974), 109; Fisher, *Hard Evidence*, 66-67.

41. O'Hara, *Fundamentals*, 582.

42. Midkiff, "Arson and Explosive Investigation," 240-41. See also Saferstein, *Criminalistics*, 336.

43. Midkiff, "Arson and Explosive Investigation," 241-43.

44. Ibid., 243.

45. Fisher, *Hard Evidence*, 84-85.

46. Saferstein, *Criminalistics*, 133, 338-39.

47. Midkiff, "Arson and Explosive Investigation," 243.

48. Saferstein, *Criminalistics*, 339-40.

49. Jim Dwyer et. al., *Two Seconds under the World: Terror Comes to America—The Conspiracy behind the World Trade Center Bombing* (New York: Ballantine Books, 1994), 11-17. Except as otherwise noted, information for this case study is taken from this source and from Fisher, *Hard Evidence*.

50. Dwyer et. al., *Two Seconds*, 37-58.

51. Ibid., 57, 59; Fisher, *Hard Evidence*, 87.

52. Dave Williams, quoted in Fisher, *Hard Evidence*, 87-88.

53. Ibid., 87.

54. Fisher, *Hard Evidence*, 90; Dwyer et. al., *Two Seconds*, 82.

55. Dwyer et al., *Two Seconds*, 91-93.

56. Ibid., 108-122.

57. Fisher, *Hard Evidence*, 93.

58. Ibid., 94.

59. Ibid.

60. Ibid., 95; Dwyer et al., *Two Seconds*, 26, 216-17.

61. Fisher, *Hard Evidence*, 95; Dwyer et al., *Two Seconds*, 297-332.

62. Dwyer et al., *Two Seconds*, 336-37.

63. Ibid., 363-65, 380. A sequel to the World Trade Center bombing case came in the form of a scandal involving the FBI laboratory. A whistle-blower, FBI chemist Frederic Whitehurst, made a number of allegations regarding the lab's explosives unit that triggered an investigation by the Justice Department. Inspector General Michael Bromwich faulted the lab for questionable scientific procedures and courtroom testimony in the World Trade Center and Oklahoma City federal building bombings, but a follow-up report concluded that the bureau had made progress in resolving the problems. (See "Nationline: FBI Lab," *USA Today,* June 5, 1998.)

64. Ibid., 243-93. See also *The World Almanac and Book of Facts 1994.* (Mahwah, N.J.: Funk & Wagnalls, 1993), 57, 59-60, 63.

RECOMMENDED READING

Bertsch, Wolfgang. "Was It Arson?" *Analytical Chemistry News & Features* (Sept. 1, 1996). An overview of the process used by forensic chemists to analyze fire debris.

Beveridge, A.D. "Development in the Detection and Identification of Explosive Residues." *Forensic Science Review* 4.1 (June 1992). Review of the traditional and novel methods of analyzing explosive residue.

Douglas, John, and Mark Olshaker. *Unabomber: On the Trail of America's Most Wanted Serial Killer.* New York: Pocket Books, 1996. Pretrial account of the manhunt for a serial bomber and the arrest of Theodore J. Kaczynski; includes "Unabomber" manifesto and details of sixteen bombing cases.

Dwyer, Jim, and David Kocieniewski. *Two Seconds under the World: Terror Comes to America—The Conspiracy behind the World Trade Center Bombing.* New York: Ballantine Books, 1994. Excellent account of a major bombing by a Pulitzer Prize-winning team of reporters.

Laska, Paul R. "Investigating a Bomb." *Law Enforcement Technology* (Aug. 1995). Discussion of how forensic experts investigate bombings such as those of the World Trade Center and the Oklahoma City federal building.

Mohnal, Thomas J. "Unabomb." *Crime Laboratory* 21.3 (July 1994). Brief account of the serial explosions attributed to the "Unabomber," written by an FBI official; includes diagrams of the explosive devices.

Saferstein, Richard, ed. *Forensic Science Handbook*, vols. 1 and 2. Englewood Cliffs, N.J.: Prentice-Hall, 1982, 1988. Authoritative forensic text covering most technical aspects of criminalistics including liquid chromatography, mass spectrometry, DNA, arson and explosive investigation, and alcohol in blood and breath (vol. 1); and controlled substances, capillary gas chromatography, paternity testing, microtraces, and individualization of semen stains (vol. 2).

10 PATHOLOGY

As mentioned in the introduction, the earliest forensic scientists were physicians who were called upon to give an opinion as to the cause of death in individuals. Increasingly in the United States, medical examiner programs are replacing the old system of elected coroners, who are not required to have training. The first state medical examiner system was established in 1939 in Maryland.[1] It is the task of the medical examiner, or the forensic pathologist (who has an M.D. degree), to assist in the *identification of decedent*, to determine the *time of death*, to conduct the *autopsy*, and to determine the *cause of death*—those topics comprising this chapter. The case study is the death of Marilyn Monroe.

IDENTIFICATION OF DECEDENT

Regardless of who has the ultimate legal burden, the coroner or medical examiner works with law enforcement personnel and others to help identify the bodies of deceased persons. According to one authority:

> Few forensic endeavors offer a more challenging and creative exercise than establishing identification of the living or dead—challenging because identification with medical and legal certainty often requires thinking beyond routine fingerprint and dental comparison. Identification workups are creative in developing features that are unique as presented by an individual decedent. Beyond humanitarian considerations, identification is essential to the completion and certification of official documents. The accurate identification of a decedent permits certification

of death and notification of next of kin. Only then may they proceed with the probate of wills, apply for disbursement of benefits and insurance, and begin to work through the grieving process. In the case of unidentified living persons, whether amnesia victims or abducted children without memory of who they are, or the injured and impaired, identification is necessary to re-establish their identity and their lives. Law enforcement agencies need positive identification to pick up the leads of investigation to develop suspects, establish the *corpus delicti* of homicide and reconstruct the sequence of events of a crime.[2]

To facilitate identification of a corpse, the following procedures are standard:

Physical Description. A basic *portrait parlé* of the person should be given, much as if describing a living person. Height should be recorded, along with weight, color of eyes and hair, etc. It should be kept in mind that decomposition can affect the color of the skin (which can blacken to the extent that a Caucasian may appear Negroid), the color of the hair (blond hair darkens, while red or brown hair tends to become lighter and gray), and the weight (which may be overestimated due to bloating). If the hair color is patchy or varies near the scalp, dying is a distinct possibility, and samples should be taken for lab analysis.[3]

Scars and Marks. Part of the *portrait parlé* that deserves special mention are skin markings—birthmarks, scars, and so on. Tattoos may be especially helpful since they may indicate some past experience (such as a military insignia) or clue as to lifestyle ("jailhouse" tattooing, for example), sexual preferences (such as gay motifs), or personal interests (such as a motorcycle insignia), and so on, including the decedent's initials or those of someone else. Sometimes tattooed numbers are found such as social security, military, or prisoner of war. Irregular scarring may indicate the former presence of a tattoo. In the case of indistinct tattoos, infrared photography, high-contrast photography, and computer image enhancement may be successful.[4]

Fingerprints. Fingerprinting is still the mainstay of identification techniques, and the basics of taking fingerprints from the deceased—in various states of decomposition—are described in chapter 5.[5] The new AFIS technology makes it possible for identification by fingerprints to be accomplished with increasing speed.

Photographs. The entire body should be photographed, as well as the full face and profile. The latter is particularly valuable in recording the shape of the ear—the feature of the face that is most like a fingerprint—which may be matched to an authenticated photograph. (See Recommended

Reading for Alfred Iannarelli.) Any scars or other distinctive features such as amputations should be documented in photos as well. The photographs may be shown to those who knew the deceased to provide a possible identification.[6]

Age Determination. The deceased person's apparent age may be roughly estimated by the teeth and—in the case of people under about twenty-five—the joining of the bones. X-rays provide a basis for studying the extent of cranial and epiphyseal fusions (the uniting of various bones that serve as landmarks for estimating the decedent's age).[7]

Dental Features. The pattern of dental work, including plates, bridges, and fillings, can be compared with known dental records, plaster molds, and x-rays to effect a positive identification. Fillings are highly distinctive and can serve as an effective means of individualization.[8] In one celebrated case, a piece of a jaw with three and a half teeth was found in the debris from a Greenwich Village bombing. When a possible victim was indicated, it was possible to obtain her fifteen-year-old dental x-rays that showed an identical shape of the roots and bone structure.[9]

Radiological Evidence. The presence of old bone fractures, shown in x-rays of the body, sometimes provide a positive identification. So may surgical pins, plates, pacemakers, and other implants.

Blood Factors. Not only do the ABO blood grouping, Rh factor, and other blood characteristics provide an additional means of identification, but DNA testing makes the blood evidence particularly valuable. In the case of deliberately mutilated remains or where there is a question (as in the case of an explosion or airline disaster) of whether body parts are from one or more persons, DNA typing can resolve the matter. DNA can also be helpful in instances of badly decomposed or skeletonized bodies (discussed further in the next chapter).[10]

Medical Indications. A thorough postmortem examination may reveal various diseases or conditions that have identification value. Among these are hypertension, old strokes, diabetes, Alzheimer's disease, and drug abuse. The presence of anticonvulsants indicates a seizure disorder; antidepressants suggest depression, possibly even suicidal circumstances; and so on.[11]

Other Means. Clues to a dead person's identity may be found at the scene where the body is discovered or may be among the personal effects secured by the police or pathologist. A driver's license or other identification card in the victim's wallet or purse may be particularly helpful, although mere possession of such identification is not conclusive. Additional clues may be provided by monogrammed personal items, such as a cigarette lighter, or by clothing. The latter may be traceable as to its purchase or it may contain laundry or dry cleaner's marks.[12]

TIME OF DEATH

As soon as possible at the site where a body is discovered, the pathologist or medical examiner should make the observations that can indicate an approximate time of death. "Hours later," one authority says of the body, "after it has been photographed, put into a body bag, transported to a morgue and put into a cooler, the postmortem changes will be different and more advanced. The sooner after death a body is examined, the more accurate is the estimation of time of death."[13]

Generally, time of death is estimated from certain changes that occur in the body following death. These changes can also provide indications of alterations in the body's position after death and help indicate whether the death was murder or suicide. The changes include *temperature*, postmortem *lividity*, *rigor mortis*, and *putrefaction*. *Stomach contents* and *ocular changes* may also be helpful in determining the time of death.

Temperature. The normal body temperature is 98.6 degrees Fahrenheit. After death the body tends to cool. The rate of cooling depending on several factors. First, the rate is affected by the temperature of the air, the cooling obviously being faster on cold days. The way the body is clothed is another factor, with heavy clothing naturally retarding the cooling. Another consideration is the amount of subcutaneous fat on the body. Fat people tend to cool more slowly, whereas aged persons, who tend to have less fat, can be expected to cool more rapidly. Apart from these factors, the temperature of a dead body averages a drop of approximately one and one-half degree Fahrenheit per hour.[14]

Lividity. The dark purplish-blue discoloration that is seen on the portions of the body that are nearest the ground is called postmortem lividity. It is caused by blood settling into the body's lower parts. For example, the forearms and lower legs will exhibit the characteristic darkening in the case of a hanging, whereas a person lying on his or her back will have lividity on the backside. If an investigator encounters, for example, a corpse lying on its back, with lividity on the *front* of the body, then the body has been moved from its original face-down position. Lividity appears about two hours after death. The blood tends to clot in the tissues after this.[15]

Rigor Mortis. Immediately after death, the body is limp due to relaxation of the muscles. (The relaxation of the sphincter muscles, for example, causes incontinence.) Over time, however, the muscles begin

FIGURE 10.1. Forensic pathologists often go to the crime scene in order to view it firsthand. Here a medical examiner and other law enforcement personnel are preparing to remove a body from the scene. Note that the hands are being "bagged" to prevent the loss of potentially important trace evidence. (Courtesy of Don Ostermeyer, J.L. Bunker & Associates, Ocoee, Florida.)

to stiffen due to chemical changes within the muscle tissue. (Accumulating waste products cause myocin in the muscles to coagulate.) This process, called rigor mortis, begins at the lower jaw and neck and spreads downward. Both the voluntary and involuntary muscles—including the heart—contract. The time when rigor sets in may be as soon as fifteen minutes following death or as long as fifteen hours after. The average time is approximately five to six hours after, with the upper portion of the body affected after about twelve hours and the entire body within some eighteen hours.[16]

After a time, rigor mortis disappears, usually within thirty-six hours. Again, the process begins with the head and neck and proceeds to the lower extremities. The disappearance may take from about eight to ten hours. The variables that affect the speed of rigor's onset and departure include great heat and individual differences in muscular development. The following rough rules may generally be employed:

1. Rigor mortis should begin within ten hours after death.

2. The whole body should be stiff within twelve to eighteen hours after death.

3. Stiffening disappears within thirty-six hours after death.[17]

It should be noted that "Under certain conditions the stiffening of the hands or arms may take place immediately at the time of death."[18] This stiffening, known as *cadaveric spasm,* is frequently confused with rigor mortis.[19] It typically occurs in cases where great tension or excitement precedes the death. In certain instances of the phenomenon, a drowning victim will have weeds or other aquatic matter clenched in the hands, or a suicide may be tightly clutching a pistol, or yet again a homicide victim may be grasping some of the assailant's clothing or hair.[20]

Putrefaction. Decomposition changes in the dead body that are caused by the action of microorganisms such as bacteria are known as putrefaction. It is characterized by bloating due to gas (which escapes with time), darkening of the face with decomposition liquids escaping from the nose and mouth, swelling of the tongue, green discoloration of the abdominal skin, and the formation of fluid- or gas-filled blisters. Putrefaction may set in quickly, as fast as a day in a tropical climate, or slowly, being scarcely observable in freezing temperatures even after months.

Further destruction of the body is caused by maggots and various insects. Flies may deposit eggs between the eyelids or lips or in the nostrils within a matter of minutes, and maggots may develop within twenty-four hours. If the temperature is above forty degrees Fahrenheit, various insects will feed upon the body until it is skeletonized.[21]

Stomach Contents. Time of death may also be approximated from the appearance and the amount of stomach contents. The determination depends on the fundamental assumption that the stomach empties at a certain, known rate. Unfortunately, the emptying rate changes depending on a number of factors. These include the type and amount of food ingested, any intake of drugs or medicine, the emotional and medical condition of the decedent, and other variables. Generally, a light meal will be in a person's stomach for up to two hours, a medium meal for up to about three or four hours, and a very heavy meal up to four to six hours or even more.[22] Therefore, if the stomach is completely empty, death probably occurred a minimum of four to six hours previously; if the small intestine is empty as well, death probably occurred twelve or more hours earlier.[23]

Ocular Changes. Some of the earliest postmortem changes may occur in the eyes. If the eyes stay open, a thin film may form on the corneal surface in a few minutes, and cloudiness may develop in two to three hours. Should the eyes remain closed, the formation of the corneal film may not occur for hours, and the cloudiness may take twenty-four hours or longer to appear.[24]

It is important to remember that time of death can only be estimated, and that any estimate is subject to error based on environmental and other variables.

AUTOPSY

The term *postmortem examination* refers to all the procedures followed by the coroner, forensic pathologist, or other qualified individual in conducting a death investigation. It includes the preliminary examination, identification of the body, photography, x-raying, etc.—and the autopsy. The *postmortem* is warranted and should be conducted in any death where homicide is suspected, including suicide. A *preliminary* examination should be conducted by a physician in deaths by criminal violence, suicides, accidental deaths, deaths where no physician was present, sudden deaths of persons in seeming good health, prison deaths, and any deaths that occur in an unusual or suspicious manner. When the preliminary examination and investigation fail to establish a clear cause of death, an autopsy should be conducted.[25] A general or medical autopsy is performed in a hospital for the purpose of discovering any pathological (disease) processes in addition to learning the cause of death. In contrast, a medical-legal autopsy is much more complete and requires special training. It includes identifying and tagging the corpse; photographing the body, both dressed and nude, with full-face and profile portraits; recording height and weight and taking x-rays; examining the external body; providing a detailed description of any bruises, ligature marks, gunshot or stab wounds, etc.; dissecting and examining the internal body; laboratory testing of organs and body fluids for drugs and poisons; and rendering an opinion, adding "cause of death" to the death certificate.[26]

In the actual dissection of the body, the thoraco-abdominal (chest-belly) cavity is opened with an incision. This is most often Y-shaped, beginning at each armpit area and running beneath the breasts to the lower end of the breastbone, then continuing downward through the middle of the abdomen to the pubis. At this point, the front portion of the ribs and breastbone is removed as a unit, exposing most of the organs for examination. Next the heart and lungs, together with the trachea and esophagus, are taken out—sequentially or, more often, *en bloc*. The abdomen is given a general examination prior to removal of the organs. Fluids are aspirated (drawn out) so that they can be analyzed. Each organ is then weighed, examined externally, and sectioned for internal study. Any fluid in what is called the thoracic pleural cavity is aspirated for later analysis. Tissues from the organs are mounted on microscopic slides for subsequent study of any cellular changes. The contents of the stomach are measured, recorded, and sampled for toxicological analysis.

In the pelvic area, the genitalia are examined with regard to any injury or foreign matter. (Vaginal and anal swabs are taken during the external examination.) Blood, semen, and hair samples are collected for lab analysis. The urine is collected following removal of the bladder so that certain drugs that tend to concentrate there (like Valium and barbiturates) may be detected by the toxicologist.

Finally, the head is examined, beginning with the eyes. To remove the brain, an incision is made across the top of the head, the scalp is pulled forward, and the skull is exposed. A saw is then used to cut through the top of the skull so that the brain may be examined, removed (after various nerves, blood vessels, and other attachments are cut), weighed, and sectioned for later microscopic review.[27]

Following the dissection of the body, the large incisions are sewn shut. Microscopic specimens are then studied, and chemical analyses are conducted in the laboratory. This completes the autopsy, after which the medical examiner synthesizes the findings and attempts to determine a "cause of death" along with any contributing factors.[28] These results are presented as a formal "autopsy protocol," a legal report and file, typically a folder that includes photographs, x-rays, fingerprints, and toxicological test results. The report includes the following:

1. External Examination:
   a. description of clothing
   b. description and identification of the body
2. Evidence of Injury:
   a. external
   b. internal
3. Central Nervous System (head and brain)
4. Internal Examination of Chest, Abdomen, and Pelvis
5. Toxicology Test Findings
6. Opinion[29]

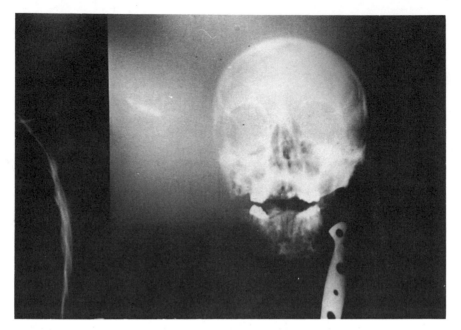

FIGURE 10.2. X-rays are vital to the work of the forensic pathologist, frequently playing a crucial role in determining the cause of death. (Courtesy of Don Ostermeyer, J.L. Bunker & Associates, Ocoee, Florida.)

CAUSE OF DEATH

For investigative purposes, it is important to distinguish between the cause, manner, and mode of death. *Cause of death* is understood to be the pathological condition that produced the death (such as subdural hemorrhage), *manner of death* refers to the physical agent or instrument that was employed (such as a blunt-force injury to the skull), and the *mode of death* indicates the intent (or lack thereof) when the instrument was used, and by whom (for example, homicide by person or persons unknown). Or, to consider another possible scenario, the *cause* may be myocardial infarct ("cardiac arrest"), the *manner* arteriosclerosis ("hardening of the arteries"), and the *mode* natural.[30]

The four modes of death are natural, accidental, homicidal, and suicidal. The findings at autopsy are often critical in determining whether a given death was caused by homicide. For example, the death of an elderly person that on the face of it appears to be due to natural causes may in fact prove to have been caused by poison. On the other hand, an autopsy may reveal that a person died of a close-proximity gunshot wound to the head without indicating whether it was caused by homicide, sui-

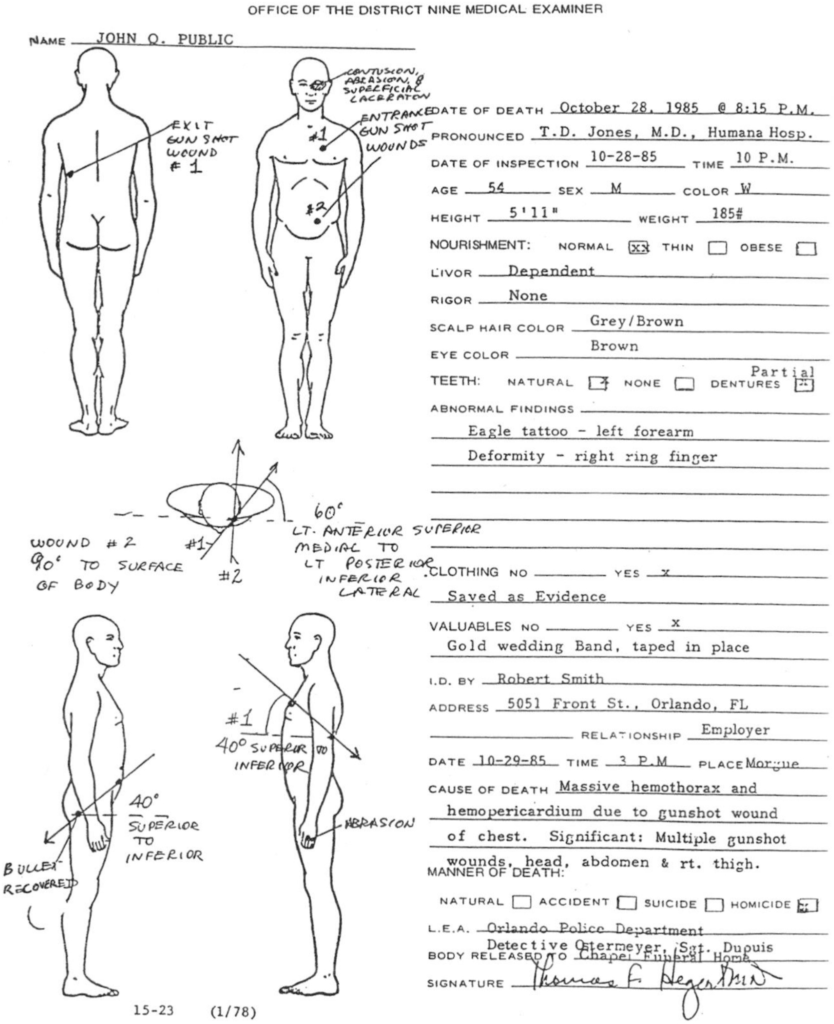

FIGURE 10.3. Autopsy notes and wound-location diagrams are generally made on a standardized form such as this. (Courtesy of Don Ostermeyer, J.L. Bunker & Associates, Ocoee, Florida.)

cide, or accident. In such an instance, the evidence from the scene and the discoveries of the investigator may prove decisive. For example, the absence of a firearm at the scene would suggest homicide, while the presence of a suicide note would, if proved genuine, be an indication of suicide. (Increasingly, a "psychological autopsy"—an investigation of a decedent's mental state prior to death—is conducted as an aid to determining whether the person committed suicide.)[31]

Given the number of means by which death can occur and the attendant complexities, we will review here only the more common methods and discuss briefly how they may be identified or distinguished from other methods. We will consider the various types of asphyxia, blunt-force injuries, burning deaths, electric shock, gunshot wounds, motor vehicle fatalities, poisoning, and stabbing.[32]

Asphyxia. Death caused by lack of oxygen is known as asphyxia. In *suffocation,* the passage of air is blocked. This may be accomplished by a pillow placed over the face, for example, with possible resulting bruises and abrasions on the inside of the victim's upper lip. Suffocation also may be accidental, with foreign material such as food obstructing the airway. In this event, the foreign material should be revealed during the autopsy.

Deaths by strangulation, when they are homicidal, may be indicated by the presence of fingertip bruises on the neck. Fingernail marks may also be present in such cases of manual strangulation. Ligature strangulation will be indicated by the presence of a groove or other marking on the victim's throat. At autopsy, strangulation may be indicated by fractures of the larynx, including the hyoid bone.

In the case of asphyxia by *hanging,* signs of violence, particularly about the neck, may be an indication of prior strangulation. If the body is found not to be completely suspended, the death is probably one of suicide, since a murderer would strive to completely suspend the body. Hanging is a common means of suicide but is relatively rare as a form of homicide or accidental death (except in the case of children).

Drowning deaths are also due to asphyxia. States one authority: "The lungs ordinarily are not filled with water. In the process of drowning, the person takes some water in the mouth and begins to choke. Irritation of the mucous membranes results in the formation of a great deal of mucus in the throat and windpipe. Efforts to breathe produce a sticky foam which may be mixed with vomit. The foam prevents the passage of air into the lungs."[33] Drowning may be indicated by the presence of this white foam. Other indicators are articles such as seaweed found grasped in the hand (an indication of cadaveric spasm) and swelling of the lungs.

Asphyxia may also be caused by a blow, such as a karate chop to the neck, or by compression of the chest, as in being pinned under a collapsed building. A cherry-red lividity may indicate asphyxia by carbon monoxide poisoning, and this or other type of poison-gas asphyxia may be determined by autopsy.

Blunt-force Injuries. These are usually directed at the head, but internal organs of the body may suffer blunt-force trauma without any external indications of violence. These may be revealed at autopsy. Blows to the head will reveal fracture of the skull; those to the chest may produce broken ribs that pierce the lungs or heart, or cause the heart to be crushed or ruptured; and blows to the abdomen may produce a ruptured spleen, liver, or kidney and may result in death from hemorrhage.

Burning Deaths. Deaths by fire typically cause the corpse to assume a distorted position somewhat resembling a boxer's pose—hence the term "pugilistic attitude." It is sometimes thought that this position is an indication that the body was alive at the time it burned, but since the effect is simply that of contracted muscles, it occurs whether the person was dead or alive. If an individual was indeed alive at the time of a fire, there will be smoke stains about the nostrils and the presence of carbon monoxide in the blood. In the case of skin that is burned without attendant singeing of the body hair or clothing, scalding by hot liquid or steam is indicated.

Electric Shock. Death by electrocution does not ordinarily produce a characteristic appearance. However, pulmonary edema (swelling of the heart) and the appearance of asphyxia may be present. High-voltage electrocution is likely to produce visible burns on the body, while one-third to one-half of low-voltage electrocutions leave no burns.

Gunshot Wounds. Bullet wounds are of two basic types: (1) the *entrance wound,* which tends to be smaller by comparison than the exit wound and also rounder and neater with a black ring of discharge products around the edges and comparative lack of bleeding, and (2) the *exit wound,* which is generally larger and more ragged in appearance and has shredded tissue extruding from the wound and much more profuse bleeding. "In the absence of eyewitnesses," states a homicide investigation textbook, "it is not always a simple matter to determine whether a death from gunshot wounds is an accident, suicide, or murder. The circumstantial evidence is helpful in some cases. Often, however, no conclusion can be drawn."[34]

The location of the wound is important. In murder it may be found anywhere, but in suicide the tendency is for the wound to be in the right temple, the mouth, beneath the chin, the center of the forehead, the center of the back of the head, or the left chest. The majority are in the right temple with the gun held tightly against the skin so as to produce the typically star-shaped "contact" wound. The presence of the weapon tightly gripped in the hand (by cadaveric spasm) is a strong indication of suicide.

The appearance of a gunshot wound can be affected by a number of factors, including the firing distance, type of weapon and ammunition,

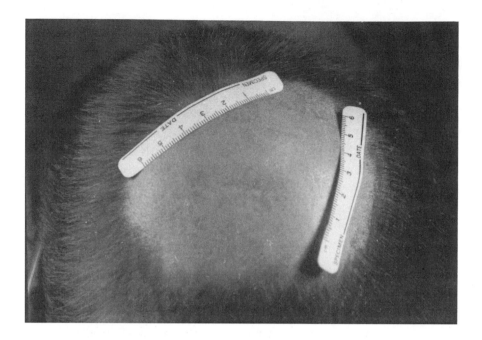

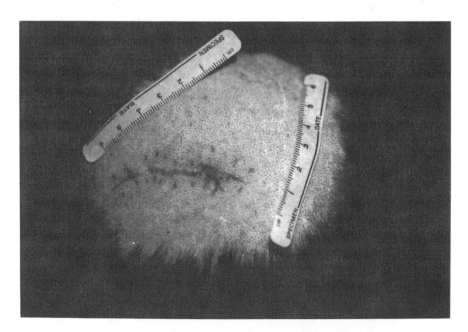

FIGURES 10.4 AND 10.5. Specialized photography is frequently used to "visualize" wound patterns. In figure 10.4, an eight-month-old wound is shown in room light. The same wound area is shown in figure 10.5, a photograph taken using ultraviolet light and a special lens and filter.

passage through clothing, part of the body affected, and other factors such as ricocheting. In many instances, some indication of the distance between the firearm and the wound is given by the presence of "tattooing" or "stippling" (pinpoint hemorrhages caused by the discharge of burned gunpowder), which indicates a relatively "close shot."

Motor Vehicle Fatalities. Traffic accidents present a number of special circumstances. In the case of a hit-and-run accident, the clothing of the victim should be scrutinized carefully for the presence of grease from the vehicle's understructure, tire marks, and fragments of glass or paint. The leg bones of the victim are frequently fractured by bumper contact, and the height of such injuries in relation to the soles of the feet should be measured for comparison with the forwardmost part of the bumper.

Trauma associated with the *driving* of vehicles that have suffered an impact may be severe. Impacts from the front often transversely fracture the sternum (breastbone) and cause rib fractures on the sides of the chest. In young adults, however, the ribs and sternum may be unbroken, due to their relative elasticity, while there are crushing injuries to the lungs and heart.

Poisoning. As we saw in the previous chapter, the positive identification of a poison is the domain of the toxicologist. But it is the pathologist who, working with investigators, develops evidence that this hidden method of death was applied in a given case. And it is the pathologist who makes the final legal determination of the cause of death in such cases.

According to Geberth's *Practical Homicide Investigation*, "Practically speaking, murder by poisoning is extremely rare." As Geberth explains: "Investigators are usually confronted with cases in which the victim has committed suicide by taking an overdose of pills or ingesting something dangerous in order to cause death. Other cases of poisoning are usually accidental and involve narcotic overdoses or the inadvertent taking of the wrong medication."[35]

Of the variety of poisons mentioned in chapter 9 (see Table 9.2), only a few are commonly used for murder. For example, murder mystery stories notwithstanding, hydrocyanic acid ("prussic acid") so quickly produces symptoms as to discourage its use for murder.[36] Of the poisons used for murder, two of the most common are the metallic poisons, arsenic and antimony. Indeed, arsenic "is the most commonly occurring poison in cases of murder."[37] Moreover, it is among the most available and accessible, particularly in the form of insecticides (used for spraying trees), rat poison, and even certain medicinal preparations. Symptoms of poisoning by arsenic include vomiting, cramps, and diarrhea, possibly

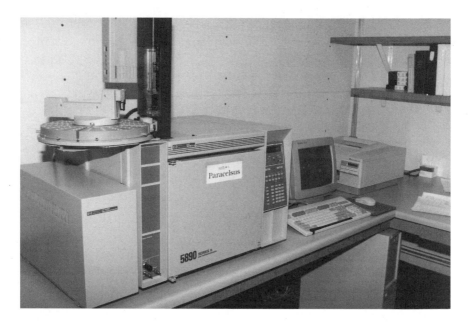

FIGURE 10.6. Gas chromatography represents a major technique for analyzing toxicological samples. Here a gas chromatograph is equipped with a "carousel" for testing numerous specimens, such as those being analyzed for blood-alcohol levels. (Courtesy of Robert Ruttman, Mesa Police Department, Mesa, Arizona.)

bloody diarrhea. Death may occur in a few hours or in several days. Sometimes murderers avoid administering a large, fatal dose, which calls attention to itself, and instead induce poisoning in a chronic, progressive manner. Arsenic may be detected in the body after many years, but the mere presence of the poison is not a conclusive indication of murder, since there are many ways traces may be ingested or even diffused from the soil in which the burial occurs.[38]

Less commonly used as a poison, antimony has various industrial and medicinal (including veterinary) applications. The symptoms of antimony poisoning include intensive gastric irritation and pain, a metallic taste in the mouth, vomiting of bloody material, diarrhea, sweating, rapid pulse, and muscle spasms. In fatal cases there may be delirium, subnormal temperature, and collapse. Poisoning may be acute (sudden) or chronic (incremental). As to the latter, according to *Legal Medicine*, "In a few cases of homicide in which the victims were invalids, the antimony compound was administered in this manner so that the symptoms of poisoning would simulate the clinical appearance of a natural disease condition."[39]

Stabbing. Stabbing and cutting wounds may cause death themselves or lead to complications such as tetanus or pneumonia that represent a secondary cause of death. Rarely are cutting or stabbing deaths accidental; usually they are the result of suicide or homicide. In suicide, the left wrist, left chest, throat, and femoral artery are the most commonly attacked locations. Sometimes the wounds are accompanied (actually preceded) by superficial cuts referred to as "hesitation marks" (indicating tests before the person summons sufficient nerve to inflict the fatal slash.) Their presence is a strong indication of suicide, as is the weapon being clenched in the deceased person's hands, indicating cadaveric spasm. In suicides the body remains at the site of the fatal cutting or stabbing.[40]

In the case of homicide, the body may have been moved or there may be a blood trail caused by the victim's attempting to flee the assailant. Whereas suicide may be accomplished by cutting only, murder tends to invite stabbing strokes as a means of efficient killing. The fatal wounds from homicide usually occur in the neck or upper chest, and such wounds in the back are obviously indicative of murder. Sometimes the victim receives cuts on the palms or outer forearms that result from attempts to ward off the attack; such wounds are known as "defensive wounds." An examination of the wounds by the medical examiner may permit an estimate as to the type of weapon and its size, shape, and sharpness. The wound dimensions and depth of penetration are useful indications of the weapon's characteristics.[41]

CASE STUDY: THE DEATH OF MARILYN MONROE

Born Norma Jean Mortenson in Los Angeles on June 1, 1926, actress Marilyn Monroe began life as an illegitimate child. She later took her mother's maiden name, Baker. After her mother was institutionalized for mental illness, Norma Jean was raised by foster parents and, at the age of only sixteen, married an aircraft worker named James Dougherty. Four years later, in 1946, she was divorced.[42]

Marilyn Monroe then began her professional career as a photographer's model, posing for nude pinups. Aided by twin assets—"uncommon beauty and driving ambition"[43]—she soon was at the brink of Hollywood stardom. In 1949, she obtained bit parts in two movies, *Ladies of the Chorus* and *Love Happy*. She won a feature role the following year in *The Asphalt Jungle*. Her career included starring roles in twenty-two additional films that grossed a total of $200 million. These included *All About Eve* (1950) and *Love Nest* (1951). By that time, Monroe was a full-fledged star. In

1952, she made five feature films: *Clash by Night, We're Not Married, Don't Bother to Knock, Monkey Business,* and *O. Henry's Full House*. These were followed by such hits as *Gentlemen Prefer Blondes* (1953), *The Seven Year Itch* (1955), *Some Like It Hot* (1959) and—her last film, starring Clark Gable—*The Misfits* (1961), which was a box-office disappointment.

In the meantime, her personal life was coming apart. Her second marriage, to baseball great Joe DiMaggio in 1954, lasted only nine months. In 1956, she married again, this time wedding celebrated playwright Arthur Miller. In just four years, that marriage too ended in divorce. By then, Monroe had become dependent on various drugs and had suffered a miscarriage. On at least two occasions she attempted suicide.

Nevertheless, the world was stunned when, on Sunday, August 5, 1962, it was reported that Marilyn Monroe was dead. A police commander, when logging an early-morning call from the medical examiner reporting the death, thought it was a joke and went personally to investigate. Even the pathologist assigned to the case, Dr. Thomas Noguchi, thought the deceased was another woman of the same name—perhaps, as he later explained, "because to me, as to almost all Americans, Marilyn Monroe was a phantom goddess of the screen, not a real person."[44]

Noguchi read the preliminary investigatory report. Marilyn Monroe had been pronounced dead by her physician, Dr. Hyman Engelberg. Several bottles of drugs, including an empty bottle of Nembutal sleeping pills and a partially empty container of chloral hydrate, were on the table next to her bed. The report noted that Dr. Engelberg "had given refill on Nembutal day before yesterday" and that "Psychiatrist talked to her yesterday, very despondent."[45] Noguchi searched for needle marks and found none, noting that fact on the standard body diagram of the autopsy report. Although he found no visual traces of pills in the stomach, the subsequent toxicologist's report showed that the blood contained 8.0 mg. percent of chloral hydrate while the liver had 13.0 mg. percent of pentobarbital (Nembutal)—both being well above fatal levels.[46]

Marilyn Monroe had scarcely been buried when the usual conspiracy vultures began circling. Most of their "theories" involved Monroe's murder being the result of a diary (apparently nonexistent) that somehow threatened Robert Kennedy. One source posited that Kennedy himself was the murderer; the fact that he was probably elsewhere became, naturally, part of a "cover-up." Other conspiracy advocates suspected rogue elements of the CIA, who also were fearful about information in the "diary." The various scenarios depended on the tragic star having died by injection rather than having taken pills, and the autopsy report was cited as evidence of the scenario.[47] Several other points were raised, each of which is easily refuted or countered.

For example, proponents of the conspiracy theory point to the contradiction between the finding of a death by overdose of drugs and the absence of pills or capsules in Marilyn Monroe's stomach or small intestines. In fact, according to famed pathologist Dr. Michael Baden: "The process of absorbing liquids into the bloodstream works faster in some people than in others. Alcohol, barbiturates, and water could all have been absorbed after Monroe lost consciousness and before she died. It is possible," adds Baden, "that the toxicologists did not examine the stomach contents as carefully as they might. They were not as sophisticated then as they are now and it's difficult to say what they might have missed."[48] But what, others have asked, about the yellow dye from the jackets of the Nembutal capsules? Shouldn't there at least have been a yellow stain in the linings of the throat, esophagus, stomach, and small intestines? The answer is no, and the question is a bogus one: The yellow color of the Nembutal capsule is made from a dye that does not run when swallowed.[49]

Another point concerns the hypothetical needle mark. Noguchi found no such mark, yet it is known that Monroe's psychiatrist had given her an injection the day before her death. If Noguchi searched thoroughly, why did he not find the medical injection? Might he have missed another, lethal one? The answer is that punctures from fine surgical needles heal within a few hours, and one made forty-eight hours previously could not have been detected. On the other hand, an injection made just before death probably would have been detected—especially by Noguchi, who discovered needle marks in the arms of John Belushi when a member of the coroner's staff failed to find any.[50] In fact, Noguchi used a magnifying lens to go over Monroe's body very carefully—so carefully that the search was praised in a report by a deputy district attorney who was present.[51]

Still another issue raised by conspiracy buffs is a bruise that was present on the movie star's body. While mystery mongers can make such a mark seem ominous, it was, in fact, merely a fresh, unexplained marking. Noguchi observed that it was of a "slight size" that "ruled against violence" and that its location was consistent with such a normal incident as "bumping into a table." He stated, "I would have expected to find fresh bruises around the throat or skull if Monroe had been a victim of violence."[52]

More significant than a small bruise, whose evidential value is virtually nil, is another finding at autopsy that confirmed that the drugs had indeed been ingested. With a massive dose of barbiturates, there is often "a corrosive or raw, red appearance on the stomach lining."[53] At least one journalist, citing in turn a medical examiner, has claimed such an appearance was absent in the case of Marilyn Monroe, but in fact it is specifically mentioned in Dr. Noguchi's autopsy report. The report stated that the "mucosa shows petechial hemorrhage diffusely." Translation: Underneath the stomach's lining or mucosa, there was observed a widespread pinpoint hemorrhaging—the very raw, red appearance that would be expected in the case of a large overdose of barbiturates.[54] The irritated appearance beneath the stomach lining provides a strong link between the missing sleeping pills (indicated by the empty Nembutal bottle and the partially empty container of chloral hydrate) and the presence of lethal amounts of barbiturates in the toxicological samples: chloral hydrate in the blood and pentobarbital (Nembutal) in the liver. Obviously, the barbiturates had passed through the stomach and had therefore not been injected, a fact supportive of the conclusion that Marilyn Monroe deliberately took the pills—their sheer number, moreover, being more than could reasonably have been swallowed "accidentally."[55]

In addition to the findings at autopsy, the evidence of a "psychological autopsy" on Marilyn Monroe pointed quite clearly to suicide as the mode of death. This was carried out by a special panel of psychological experts appointed by the chief medical examiner, Dr. Theodore J. Curphey, due to the controversial nature of the case. Although the interview notes were kept secret (not as part of a coverup but as a means to ensure complete and frank disclosure on the part of witnesses), the panel concluded that Marilyn Monroe's death was a suicide.[56]

Several factors indicate Marilyn Monroe's psychological condition at the time of her death. According to one investigator:

> Not long before her death, Monroe had been dismissed from a 20th Century Fox production, *Something's Got to Give*. On the morning of her death, she reportedly asked her housekeeper, "Mrs. Murray, do we have any oxygen?"—a revealing question, since oxygen is used to resuscitate seriously ill patients. In the afternoon she quarreled with an old friend. In the evening, she spoke to actor Peter Lawford by telephone; according to Lawford, she sounded drugged and depressed. In the course of the conversation with Lawford, who was President Kennedy's brother-in-law, she said, "Say good-bye to Jack [President Kennedy], and say good-bye to yourself, because you're a nice guy." This melancholy line has been taken as the equivalent of a suicide note.[57]

The "Suicide Investigative Team" found that Marilyn Monroe had long suffered from depression and unpredictable mood swings. For many years, she had taken sedatives and knew that an overdose would be lethal:

In our investigation, we have learned that Miss Monroe had often expressed wishes to give up, to withdraw, and even to die. On more than one occasion in the past, when disappointed and depressed, she had made a suicide attempt using sedative drugs. On these occasions, she had called for help and been rescued.

From the information collected about the events of the evening of August 4th, it is our opinion that the same pattern was repeated except for the rescue. It has been our practice with similar information collected in other cases in the past to recommend a certification for such deaths as probable suicide.[58]

As to Marilyn Monroe's apparent intent to seek rescue, it should be mentioned that when her body was discovered her arm was outstretched, with her hand resting on the telephone.[59]

Finally, there is the fact that, when help was finally summoned by the housekeeper, Mrs. Murray, Marilyn Monroe's bedroom door was locked from the inside, and it was necessary for a window to be broken to effect an entry. These are conditions that do not suggest murder. Also, Mrs. Murray was in her own room, just down the hall, the entire evening, and she heard nothing that would indicate intruders or, indeed, anything of an unusual or suspicious nature.[60]

The official conclusion of Los Angeles's chief medical examiner/coroner, Dr. Curphey, based on the autopsy and the psychological autopsy, was that Marilyn Monroe had died of "a self-administered overdose of sedative drugs and that the mode of death is probable suicide."[61] Twenty years later, this conclusion was reaffirmed when Los Angeles District Attorney John Van de Kamp released a report on that office's reinvestigation into the death.

> Her murder would have required a massive, in-place conspiracy covering all of the principals at the death scene on August 4 and 5, 1962; the actual killer or killers; the Chief Medical Examiner-Coroner; the autopsy surgeon to whom the case was fortuitously assigned; and almost all of the police officers assigned to the case, as well as their superiors in the LAPD . . . our inquiries and document examination uncovered no credible evidence supporting a murder theory.[62]

The district attorney's office conducted a thorough investigation. Allegations were patiently followed up and found to lack merit insofar as the investigators were concerned. An independent expert, Dr. Boyd Stephens, chief medical examiner/coroner in San Francisco, concluded that the original 1962 autopsy was scientifically correct and that "even the application of more advanced—1982—state-of-the-art procedures would not, in all reasonable probability, change the ultimate conclusions reached by Dr. Noguchi in 1962."[63] That conclusion remains valid today.

NOTES

1. Werner U. Spitz, ed. *Spitz and Fisher's Medicolegal Investigation of Death*, 3d ed. (Springfield, Ill.: Charles C. Thomas, 1993), 4-9.

2. Marcella Farinelli Fierro, "Identification of Human Remains," in Spitz, *Medicolegal Investigation of Death*, 71.

3. Charles E. O'Hara, *Fundamentals of Criminal Investigation*, 3d ed. (Springfield, Ill.: Charles C. Thomas, 1973), 482-83.

4. Fierro, "Identification," 103, 105.

5. Joe Nickell, *Camera Clues*, (Lexington: Univ. Press of Kentucky, 1994), 93-96.

6. O'Hara, *Fundamentals*, 483.

7. Ibid.; Fierro, "Identification," 97.

8. O'Hara, *Fundamentals*, 483-84.

9. Thomas A. Gonzales et. al., *Legal Medicine: Pathology and Toxicology*, 2nd ed. (New York: Appelton-Century-Crofts, 1954), 30.

10. O'Hara, *Fundamentals*, 484-85.

11. Fierro, "Identification," 103.

12. O'Hara, *Fundamentals*, 481-82.

13. Keith D. Wilson, M.D., *Cause of Death: A Writer's Guide to Death, Murder, and Forensic Medicine* (Cincinnati, Ohio: Writer's Digest Books, 1992), 63.

14. Ibid., 67; O'Hara, *Fundamentals*, 507.

15. O'Hara, *Fundamentals*, 508. For a photograph of lividity in hanging, see Spitz, *Medicolegal Investigation of Death*, 463.

16. O'Hara, *Fundamentals*, 508.

17. Ibid., 509.

18. Vernon J. Geberth, *Practical Homicide Investigation*, 2nd ed. (Boca Raton, Fla.: CRC, 1993), 175.

19. Ibid.

20. Joshua A. Perper, "Time of Death and Changes after Death," in Spitz, *Medicolegal Investigation of Death*, 28, 31.

21. Ibid., 31-35; O'Hara, *Fundamentals*, 510-11.

22. Perper, "Time of Death," 28-31.

23. Geberth, *Practical Homicide Investigation*, 178-79.

24. Perper, "Time of Death," 23.

25. O'Hara, *Fundamentals*, 503.

26. Wilson, *Cause of Death*, 73.

27. Ibid., 78-81; Kenneth V. Iverson, M.D., *Death to Dust: What Happens to Dead Bodies?* (Tucson, Ariz.: Galen Press, 1994), 130-32.

28. Iverson, *Death to Dust,* 133.

29. Wilson, *Cause of Death,* 88-89.

30. Geberth, *Practical Homicide Investigation,* 424.

31. Ibid., 425; O'Hara, *Fundamentals,* 547. For a discussion and example of a "psychological autopsy," see Joe Nickell with John Fischer, *Mysterious Realms: Probing Paranormal, Historical, and Forensic Enigmas* (Buffalo, N.Y.: Prometheus Books, 1992), 117-20.

32. Except as otherwise noted, information for the discussions of methods of death is taken from O'Hara, *Fundamentals,* 511-80; Spitz, *Medicolegal Investigation of Death,* passim.; Geberth, *Practical Homicide Investigation,* 223-31.

33. O'Hara, *Fundamentals,* 513.

34. Ibid., 526-27.

35. Geberth, *Practical Homicide Investigation,* 246-47.

36. O'Hara, *Fundamentals,* 532.

37. Ibid.

38. Ibid.; Gonzales et al., *Legal Medicine,* 737-46; Perper, "Time of Death," 19-20.

39. Gonzales et al., *Legal Medicine,* 747.

40. O'Hara, *Fundamentals,* 522.

41. Ibid., 522-23.

42. Biographical details are taken from her obituary in the 1963 *Britannica Book of the Year* and from Thomas T. Noguchi with Joseph DiMona, *Coroner* (New York: Pocket Books, 1983), 62-68.

43. Marc Mappen, *Murder and Spies, Lovers and Lies: Settling the Great Controversies of American History* (New York: Avon Books, 1996), 220.

44. Noguchi, *Coroner,* 70.

45. Ibid., 70-71.

46. Ibid., 72-76.

47. Ibid., 77-79.

48. Michael M. Baden, *Unnatural Death: Confessions of a Medical Examiner* (New York: Random House, 1989), 3, 4.

49. Noguchi, *Coroner,* 81-82.

50. Ibid., 81, 82, 238.

51. Ibid., 72, 82, 85.

52. Ibid., 73.

53. Dr. Sidney B. Weinberg, as quoted by journalist George Carpozi in ibid., 81. Weinberg, a retired medical examiner, was also quoted by Carpozi as stating: "The evidence points to all of the classic features of a homicide, much more so than a suicide." Ibid., 79.

54. Noguchi, *Coroner,* 80-82.

55. Ibid., 87.

56. Ibid., 75.

57. Mappen, *Murder and Spies,* 223-24.

58. Ibid., 224.

59. Noguchi, *Coroner,* 69.

60. Ibid., 88.
61. Quoted in Mappen, *Murder and Spies*, 221.
62. Noguchi, *Coroner*, 84.
63. Quoted in ibid., 85.

RECOMMENDED READING

Baden, Michael M. *Unnatural Death: Confessions of a Medical Examiner*. New York: Random House, 1989. Highly readable accounts of the work of a noted forensic pathologist, including such cases as Sunny von Bulow, John Belushi, and President John F. Kennedy's assassination.

Gonzales, Thomas A., et al. *Legal Medicine: Pathology and Toxicology*, 2nd ed. New York: Appleton-Century-Crofts, 1954. A somewhat dated but nevertheless useful text, particularly for its information on drugs and poisons.

Iannarelli, Alfred V. *Ear Identification*. Fremont, Calif: Paramont, 1989. A revised edition of Iannarelli's classic text, *The Iannarelli System of Ear Identification*, the definitive textbook on the science of identifying individuals by the patterns of their external ears.

Mappen, Marc. *Murder and Spies, Lovers and Lies: Settling the Great Controversies of American History*. New York: Avon Books, 1996. A responsible attempt to resolve famous controversies such as the Lindbergh kidnapping, the case of Sacco and Vanzetti, and the death of Marilyn Monroe.

Nickell, Joe, with Joe F. Fischer, *Mysterious Realms: Probing Paranormal, Historical, and Forensic Enigmas*. Buffalo, N.Y.: Prometheus Books, 1992. Chapter 7, "The Case of the Shrinking Bullet," includes an example of a "psychological autopsy" used in a death investigation.

Noguchi, Thomas T., with Joseph DiMona. *Coroner*. New York: Pocket Books, 1983. Studies of Noguchi's famous cases, including Natalie Wood, Marilyn Monroe, Robert F. Kennedy, Sharon Tate, and John Belushi.

Spitz, Werner U., ed., *Spitz and Fisher's Medicolegal Investigation of Death*, 3d ed. Springfield, Ill.: Charles C. Thomas, 1993. Authoritative text on most aspects of forensic pathology, including identification of human remains; well illustrated.

Wilson, Keith D. *Cause of Death: A Writer's Guide to Death, Murder, and Forensic Medicine*. Cincinnati: Writer's Digest Books, 1992. Popular text intended for mystery writers but useful for the interested layperson.

11 ANTHROPOLOGY

The pathologist's domain is that of dead bodies; the forensic anthropologist applies his expertise to skeletal remains. "In between, we share," quipped one anthropologist, commenting on the interrelationship of the respective disciplines and the frequent cooperation between experts in those fields.[1]

The noted forensic anthropologist, the late William R. Maples of Gainesville, Florida, was a former president of the American Board of Forensic Anthropology who had examined the skeletal remains of such famous historic personages as President Zachary Taylor and Francisco Pizarro. In his book, *Dead Men Do Tell Tales*, Dr. Maples states:

> I do not seek out the illustrious dead to pay them court or borrow their fame. To me, the human skeleton unnamed and unfleshed is matter enough for marvel. The most fascinating case I ever had involved a modern, love-struck couple with very ordinary names: Meek and Jennings. It fell to me to extricate their bones, burned and crushed and commingled in thousands of fragments, from a single body bag, and put them back together again as best I could. When I was finished, after a year and half's work, what I had was what lies deepest within all of us, at our center; that which is the last of us ever to be cut, burned, disassembled or dissolved; that which is strongest, hardest and least destructible about us; our firmest ally, our most trustworthy companion, our longest surviving remnant after we die: our skeleton.[2]

In this discussion of the scientific study of skeletons for forensic purposes, we look at the *recovery of remains, skeletal examination,* and *forensic*

269

identification. The case study for this chapter is the assassination of the Romanovs.

RECOVERY OF REMAINS

The remains encountered by the forensic anthropologist, who has a Ph.D. in physical or biological anthropology, exist in a bewildering variety. The expert's knowledge of the animal kingdom pays frequent dividends: bones from a bear's paw may look deceptively human; sheep's ribs may likewise resemble those of a person; and certain turtle and tortoise shells have been misidentified by lay persons as pieces of human skullcaps.[3] Sometimes human and animal bones will be mixed together. Many skeletons are recovered incomplete due to dismemberment, or to animals feeding on the corpse, or to other causes. There are remains that have been boiled, sawed into pieces, or charred from accidental fires or deliberate attempts to destroy the corpse.[4]

Therefore, skeletal remains may come to the forensic anthropologist in widely diverse forms. Examples include a simple peat-encrusted skeleton found in a wooded area, a skullcap discovered by scuba divers under a dismembered girl, and even some ten thousand fragments of charred bone from two human bodies and a dog. The fragments had all been sent in a single vinyl body bag. The boiled skull had been hidden in a paint can. Recovered remains of American soldiers who had been missing in action in Vietnam are placed in wooden cases that are in turn shipped in coffin-sized aluminum cases. "Cremains" from a cremation may be placed in a funereal urn after they have been transported in a cardboard shipping container.[5]

When the skeletal remains are not sent to the anthropologist, the anthropologist must go after them. Sometimes they are conveniently sequestered in an above-ground tomb (like the body of President Taylor, which was encased in a rotted wooden casket, the lead liner of which was, however, intact).[6] Or the remains may be exhumed from a traditional below-ground burial. In any such cases of exhumation, the following general steps are recommended:

1. A court order legally permitting the exhumation
2. The exact location of the burial:
   a. the cemetery address and the name of the individual in charge
   b. The plot number of the gravesite (or, lacking such designations in the case of smaller cemeteries, whatever alternative proof of burial may be obtain-

able). It is essential that there be no error in exhuming the very person named in the court order.

3. The date and time of the exhumation
4. A complete list of the persons attending the exhumation. These are the persons immediately concerned with the death inquiry and not members of the news media or other spectators.
5. Scaled sketches of the gravesite
6. Photographs:
   a. The burial site
   b. The coffin *in situ* (i.e., before removal from the grave or vault)
   c. The coffin above ground[7]

The coffin should be transported to the morgue or other facility for careful opening because any improper handling may raise questions about any injuries to the body that may subsequently be discovered.[8]

The container in which the body was encased can have a profound effect on preservation of the remains. According to Dr. Maples:

> The burial container is terribly important. Sealed containers protecting the body from the environment, be they a sealed steel casket costing thousands of dollars or a cheap container made of plastic or Styrofoam, will result in an amazing degree of preservation, even over long periods of time. I've seen a well-embalmed body—an autopsied body, which makes embalming very difficult—last inside a sealed casket within a burial vault for twenty-seven years, looking as if death had taken place only a day or so before, with perfectly natural features and only small areas of skin slipping from the hands and feet. I have seen other bodies buried in wooden caskets that soon disintegrated, leaving the bones badly damaged, with virtually no remaining soft tissue. I had a case in which a newborn infant was wrapped in textiles, enclosed in a plastic bag, shut up in a vinyl suitcase and buried in sandy soil for ten years. When we excavated the remains we still found soft tissue preserved, keeping those tiny, delicate bones in their respective positions, preserving them as well as if they had belonged to a fresh body buried only a few weeks earlier.
>
> . . . .
>
> Even unshielded by any container, a body will last longer underground. The general rule of thumb for the rate of decomposition is: one week in the open air equals two weeks in water, equals eight weeks underground. The horrific picture of "worms" devouring a buried corpse is false. Flies will lay eggs on a body even before it is dead, and their wriggling, wormlike larvae, known as maggots, will hatch out in just under twenty-four hours. The cycle is so regular that it can sometimes be used to establish the time of death. But maggots cannot live under-

ground. My colleague, Doug Ubelaker of the Smithsonian Institution, investigated an ancient Arikara Indian burial site in South Dakota and found that fly pupal cases were present in 16.4 percent to 38.3 percent of the burials at five sites, even though the burials were over two feet deep. How did they get there? Flies and beetles do not burrow more than a few inches below the ground. The answer is, the insects found their way to the corpse before it was buried, and were buried alive with it. When we examined the remains of Zachary Taylor, we found fly pupal cases among the bones. The industrious flies of Washington, D.C., had been at work on Taylor's body as it lay in state. They are no respecters of rank.[9]

To understand the process of natural corruption, in the late 1970s forensic anthropologist William Bass created the Anthropology Research Facility (ARF) at the University of Tennessee at Knoxville. Obtaining bodies from the local medical examiner's office, Bass began exposing them to the elements in a "decay rate facility" or "open-air morgue." Some thirty to forty bodies are exposed each year—placed on the earth or in pits, automobiles, etc.—and are carefully monitored to document the process of dissolution. (It should be noted here that what is often regarded as "miraculous" preservation of a body is actually due to some physical mechanism that has retarded decomposition. In addition to embalming, such metal poisons as arsenic and antimony may enable a body to resist putrefaction if present in a sufficiently large amount.[10] Other preservative effects are postmortem changes due to environmental conditions. These include *mummification*, the result of tissues drying under concomitant conditions of high temperature, low humidity, and good ventilation; *adipocere*, the transformation of body fat to a soap-like material due to excessive moisture; and *freezing*.[11])

When the remains are sought from clandestinely buried bodies, a well-planned search is in order. A helicopter may be helpful in spotting disturbances of soil or vegetation. Aerial photography with infrared film may be helpful because the film is sensitive to heat emitted from decomposing bodies. In the case of a ground search, some form of grid system, tied to available landmarks, should be employed, with maps provided to specially designated coordinators. Graves may be indicated by a sunken area or an adjacent mound of surplus dirt. Damage to vegetation may be apparent. A standard search technique involves probing with a steel rod of about five-sixteenths of an inch in diameter and five feet in length; one end has a T-shaped handle, while the other is sharpened. Probing can detect a soft spot relative to surrounding soil that may indicate a grave. Then a special probe that is sensitive to temperature is inserted, the reading being used to calibrate another instrument—a methane-gas detector that consists of yet another special probe, linked by a rubber tube to an instrument box. Since gases from a decomposing body rise upwardly in an inverted-V shape, probings should be made at varying depths to help ensure adequate coverage.[12]

When remains are tentatively located, the procedure described in chapter 2 should be followed. That is, a grid is superimposed on the site (oriented to north), using stakes and string; the scale drawing is prepared with corresponding squares for the plane (top) view; and an elevation (side) view is added to locate objects as to depth. Dirt should be sifted as it is carefully removed, and any small objects thus discovered should be photographed in place on the screen. The general rules of collecting and preserving evidence should be followed throughout the excavation of the scene.

SKELETAL EXAMINATION

Excavated remains are cleaned in the forensic anthropologist's laboratory on steam tables installed for the purpose.[13] Next, the various bones are placed on an examination table and arranged in proper anatomical order. Examination of the skeletal remains follows in an attempt to determine the approximate age of the individual, the sex, race, height, and other characteristics—all of which may assist in the subsequent attempt to identify the remains and determine the cause of death.

Age. As indicated in the previous chapter, the epiphysis (or stage of uniting) of certain bones can be used to estimate age. These include certain "ossification centers" where calcium and other minerals are deposited to form bone. By means of a series of formulas, age can be estimated with considerable accuracy up to twenty-five years. Similarly, the state of fusion of the sutures of the skull represents a somewhat reliable means of estimating age. There are also successive changes in the pelvic bone that occur within intervals of about five years and thus provide an aid to age determination. And there are other indications of age such as calcification that begins in the rib cartilage at the age of about fifty-five and arthritic "lipping" that appears in the vertebral column and certain joints of males in the thirty-five- to forty-year range.[14]

In addition, there is also potential evidence of age in the dental data, which includes the eruption of teeth, root formation, and crown structures—evidence that "has placed the dental estimation of chronological

FIGURE 11.1. A grid is laid out over a suspected burial site to assist in the recovery and accurate documentation of discovered evidence.

age as the primary method of age determination from birth to fourteen years of age."[15]

Sex. Items of clothing and personal possessions found with the remains may provide a suggestion of the victim's sex. The most accurate skeletal indication comes from the pelvis. The inverted-V shape of the lower contour of the bone is narrower and more pointed in male skeletons, whereas that of the female is wider and rounder—a readily discernible difference to the anthropologist's trained eye. Also, there is now an ischium-pubis index, a metrical technique for such sex determination. And what are known as "scars of parturition" (childbirth) on the pelvic bone not only indicate sex but also represent evidence that the decedent has borne children.[16] When the pelvis is not available, the diameters of the heads of the humerus (upper-arm bone), radius (the bone on the thumb side of the forearm), and the femur (or upper-leg bone) provide a probable indication of sex. See table 11.1.[17]

Race. Remnants of hair may provide a quick determination of race—*if* it is clearly Negroid or Caucasian. (The hair of Hispanics, Native Americans, and Orientals may be confused, so determination should be made by a criminalist with expertise in hair.) Determination of the race of a

FIGURE 11.2. Evidence discovered at a burial site must not only be documented as to its horizontal location but the vertical direction must be shown as well, as in this crime-scene photograph.

skeleton is made from an examination of the skull, pelvis, and long bones. Generally, a Caucasian skull has a flat front profile while African Americans exhibit prognathism (the jaws project beyond the upper part of the face).[18] Sophisticated statistical methods applied to skull dimensions can now identify the race of skeletal remains with a high degree of accuracy.[19] In the absence of the skull, the long bones of the extremities provide an indication of race, although not a certain one. The white individual's pelvis tends to be broader and to have a lower symphysis (pair or pubic bones), and the long bones of the extremities tend to be longer and straighter in the case of blacks.[20]

Height and Other Characteristics. Measurements of the long bones provide an estimate of the living stature of an individual, within an accuracy level of about an inch. (An old rule of thumb is that the height *roughly* equals five times the length of the humerus.) Whether the deceased was emaciated or obese can be determined from the skeletal remains, and muscle attachments to the bones may indicate whether or not the person was muscularly well-developed. Also, whether the deceased was right- or left-handed can be learned because the dominant arm will have slightly

TABLE 11.1.
Measurement Indices for Sex Determination

HEAD DIAMETER	MALE	FEMALE
humerus	47mm or greater	43mm or less
radius*	24mm or greater	21mm or less
femur (vertical diameter)	45mm or greater	43mm or less

* Discriminates between male and female more than 90 percent of the time.

longer bones. Past traumas, particularly fractures, also may be revealed by the skeletal remains.[21]

Cause of Death. On occasion, the bones will reveal the cause of death. For example, the skull will readily yield evidence of a gunshot wound by exhibiting a telltale pattern. Bullets may also lodge in or inflict damage to other bones that is identifiable as such. Blunt-force trauma to the skull, especially, but also to other bones may be revealed by the imprints the blows have left.[22] For example, in the case of a skullcap brought to his office, which he readily identified as belonging to a mature female, Dr. Maples elaborated on the cause of death:

> I told the deputies that the woman had been struck at least twice by a weapon that had a hammerlike aspect to it. One of the fractures was a round penetration of the frontal bone, which clearly showed the circular mark of the hammer. A small portion of bone was broken at the edge but hinged downward, indicating that the bone was fresh and elastic when the injury took place. Fracture lines radiated out from this penetration. There was another, second injury, that consisted of a depressed skull fracture of the outer layer of the cranial vault. Here the outer layer had been mashed down, but again you could see clearly the flat, circular striking surface of the hammer head. A depressed skull fracture of this type is also an indication that the bone was still fresh and elastic when the blows were struck. The skull resembles an eggshell that has been cracked but not quite broken through.[23]

Knife marks also may be discovered on the ribs and elsewhere. For instance, in the case of the historical examination of Francisco Pizarro (1470?-1541), the Spanish conquistador, Maples saw that there were "no fewer than four sword thrusts to the neck." In one of these, "a double-edged weapon had entered the neck from the right side and nicked the first cervical vertebra." Among many other such injuries, "the sixth thoracic vertebra clearly showed the marks of stab wounds from a blade

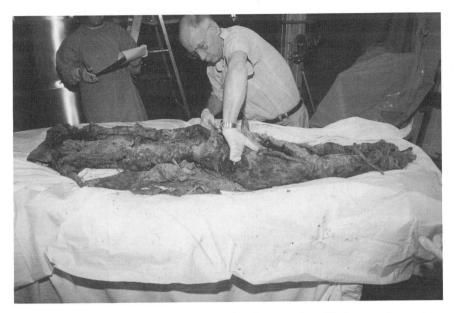

FIGURE 11.3. The late Dr. William Maples, internationally-known forensic anthropologist, examines a decomposing corpse recovered from a remote burial site.

thrust downward into the body at an angle of fifteen degrees." Also observed were a nicked rib and "defensive wounds" to the hands."[24]

Other modes of death that may be detected in skeletal remains include poisoning by one of the metallic poisons, such as arsenic, which may still be discovered in the bones after many years.[25]

FORENSIC IDENTIFICATION

The identification of skeletal remains presents particular difficulties. Here we look at three special techniques that are frequently utilized in the field of forensic anthropology. They are two standard techniques, forensic odontology and facial reconstruction, and an exciting new technological breakthrough, mitochondrial DNA.

Forensic Odontology. When fingerprints are not available, the recourse is to dental records. Because of longevity—enamel is the hardest substance of the body, outlasting all other tissue—human dentition offers obvious potential for identifying skeletal remains. Dentition also provides sufficient characteristics for individualization. The adult has thirty-two teeth, each of which presents five surfaces in a visual examination, for a total of 160 surfaces. The possible combinations of missing teeth, fillings, carious lesions (or "cavities"), and prostheses (such as bridges)—even unusual spaces between teeth—represent the basis of dental identification.[26] Forensic odontology—"the scientific application of dentistry to legal matters"[27]—includes such matters as dental identification and bite-mark comparison.

Whether involving a corpse or skeleton, the process of dental identification begins with the postmortem dental examination which should include the following (where applicable):

Missing, unerupted or extracted teeth; supernumerary teeth
Restorations and prostheses (surfaces, morphology, configuration and material)
Decay (surfaces and configuration) and broken teeth
Malposition, overlapping, crowding and spacing
Malrotation (abnormal rotation of teeth)
Peculiar shapes of teeth (e.g., peg-shaped lateral incisors)
Root canal therapy on x-ray examination
Bone pattern on x-ray examination
Complete dentures (type, shade, and material)
Relationship of the bite
Oral pathology (tori, gingival hyperplasia, etc.)[28]

The first three items listed constitute the basis of most routine dental identifications.[29]

After the postmortem examination is completed, the comparison process begins with whatever available antemortem records may be available for the proposed deceased. These should consist of all available material—dental charts, x-rays, even dental casts if possible. Of course, the more recent this material is, the more reliable and valuable it will be. X-rays are especially valuable for their ability to show the very distinctive patterns of individual restorative work. (In other words, the actual shapes of fillings are highly individualistic and offer high comparative value.) X-rays also provide information regarding areas not available to visual inspection such as root tips and the pattern of the jaw bone. Even if antemortem dental records are unavailable, an authoritative source points out that "certain noticed dental peculiarities or restorations (gaps between teeth, broken teeth, gold crowns)" as described by friends and relatives of the suspected decedent "may stand the test of comparison and, if specific enough, may enable an identification to be made." This source adds that "such information is better than none at all and may aid in the corroboration of identification by other means such as jewelry or medical data."[30] Again,

DENTAL IDENTIFICATION RECORD

Outline all caries and restorations on chart.
Shade in restorations only. See code below.

Police Case No._____
M. E. Case No._____
City_____Date_____

Circumstances requiring examination:

_____
_____
_____
_____

Site of examination:

_____
_____

Jaw Relationship
☐ Normal
☐ Protruding upper jaw
☐ Protruding lower jaw

A	Amalgam filling
G	Gold filling
S	Silicate or plastic filling
X	Tooth missing
LPM	Lost postmortem
PC	Porcelain crown
GC	Gold crown
SC	Steel crown
VC	Veneer crown
B	Fixed bridge (use brackets beginning and ending with supporting teeth)
PD	Partial denture
FD	Full denture

Prosthetic Appliances
  Maxilla
    ☐ Fixed Bridge
    ☐ Partial Denture
    ☐ Full Denture
  Mandible
    ☐ Fixed Bridge
    ☐ Partial Denture
    ☐ Full Denture
Describe prosthetic appliances:

_____
_____
_____
_____
_____
_____

This box to be filled in if identification is
made.   Date_____   Time_____

Name of Deceased_____
Street and No._____
City_____State_____
Birthdate_____Age_____
Sex_____Race_____

Abnormalities and other oral conditions:

_____
_____
_____
_____

Postmortem dental photographs   Yes ☐ No ☐
Postmortem dental x-rays   Yes ☐ No ☐

Name and Address of Examiner:

_____
_____
_____

Antemortem records obtained from
Name_____
Street and No._____
City_____
State_____
Area Code_____Tel._____
Describe antemortem records:

_____
Signature of Examiner

_____
Name of Investigating Officer

SP-144-C Rev. Sep., 1974

Additional notes on back

	Yes	No
Additional notes on back	☐	☐

FIGURE 11.4. Dental records may be essential in identifying a victim whose fingerprints have been destroyed by decomposition or which were never recorded. A typical dental identification record form is shown here. (Courtesy of Don Ostermeyer, J.L. Bunker & Associates, Ocoee, Florida.)

In terms of probability, the more dental work performed in a given individual, the greater the number of surfaces altered and hence the higher the available points of comparison. Indeed children with few or no fillings may represent difficult identification problems. On the other hand, even a single filling, if specific enough in location, morphology and material, may be sufficient to establish identity. At the other extreme of age, elderly individuals with upper and lower dentures also may present an identification dilemma. Even complete dentures, however, may retain the signatures of the attending dentist and his laboratory. This is especially true of older dentists who still personally construct their own denture cases. These dentists often can recognize their work at a glance, much like other artisans with their handicrafts.[31]

The forensic odontologist is also concerned with bite-mark injuries, which may appear on the body of a homicide victim. Bite marks are relatively distinctive and may be individualized to a particular suspect.[32] The notorious serial killer Ted Bundy was convicted in part on the evidence of a bite mark on one of his victims' left buttock. Fortunately, the mark had been photographed, with a ruler included for scale. At Bundy's trial, forensic dentist Richard Souviron used enlarged photos of the bite mark and of Bundy's teeth with the lips pulled back to show how an acetate overlay of the teeth exactly fit the bite mark.[33])

Facial Reconstruction. Since the early nineteenth century, artists have attempted to reconstruct facial appearances from skulls. Today the procedure is a controversial one among forensic anthropologists, but in some cases—often as a last resort—a reconstruction of the deceased person's face may be useful. By circulating photographs of the reconstruction, much as one would do with a police-artist sketch or a composite picture of a suspect, people who knew the victim may recognize the reconstruction and come forward with an identification.[34]

Facial reconstruction begins with an examination of the remains to determine race, sex, and approximate age of the deceased, as well as whether the person was thin, muscular, or fat. The average thickness of soft tissue over certain points on the skull is known from extensive compiled data and is modified in light of the information gleaned from the examination. At the various points, small cylinders of the appropriate thicknesses are glued in place. These are next connected by strips of clay, and the whole is filled in and shaped until a complete sculpted face is produced. A wig is selected—whenever possible, based on hair found with the remains.

The resulting facial reconstruction may then be photographed and the photos distributed by the appropriate law enforcement agency. Even a tentative identification will permit authorities to obtain the necessary dental records or other evidence that may provide positive identification. To facilitate recognition, the reconstruction photos are typically accompanied by a detailed description such as one used by the Mississippi State Crime Laboratory and divided into the following sections: Background; Time of Death; Age at Death; Sex; Race; Stature; Physique; Handedness; Parity (i.e., the condition of being parous, or having given birth to children); Old Diseases, Injuries, Anomalies; Trauma; Cause of Death; Manner of Death; and Summary.[35] One such summary read:

> This skeleton is that of a white female who was around 40 ± 6 years of age at the time of her death from unknown causes. Location of the body and binding of the hands and feet indicate homicide as the manner of death. The victim was about 5'4" ± 3" in height and was somewhat more muscular than the average female. She was right-handed and had suffered a fractured nose and broken right collar bone some years prior to her death. Death is estimated to have occurred about 2-6 months prior to the discovery of her skeleton.[36]

At times, when facial features are present but are too distorted or gruesome for public dissemination, the forensic artist may make a sketch that removes distortion, adds missing areas, and in general makes the whole presentable. Alternately, techniques like Identi-Kit and Photo-Fit utilize various facial features that may be assembled in different combinations to make composite portraits.[37]

When photographs of the proposed deceased become available, they can be compared with the skull by means of photographic superimposition.[38] For example, in the case of the infamous Nazi "Angel of Death," Dr. Joseph Mengele, whose skeletal remains were discovered in Brazil in 1985, a process known as "electronic supraposition" was employed. Two video cameras were used with a video mixer to superimpose the remains of the skull onto an actual-size photo of Mengele's head. This resulted in a good match.[39]

Mitochondrial DNA. Dramatic new technology has come to the aid of the beleaguered forensic anthropologist who is left with so little information on which to base an identification. The standard type of DNA, called nuclear or genomic DNA, is found in cells having nuclei, and it is rapidly lost in the decomposition of human remains or in exposure to heat. Fortunately there is another type of DNA that is found not in the nucleus but throughout the cell. According to one source, "changes in mitochondrial DNA are extremely rare, and happen on the order of once every three to four thousand years. That is the wonderful thing about mitochondrial DNA. It stays the same in a family for generation after generation and is passed on through the female line. It can endure in our bones for hundreds of years, if they are not cremated."[40]

The reason mitochondrial DNA is passed only through the female line should be mentioned. While it is found in the egg, it is not carried by the sperm except in the shaft or tail. This breaks off at the time the sperm fertilizes the egg at conception, so the male mitochondrial DNA is lost, and only that of the mother is passed on. It will be found in the blood, for example, of all of her children, both females and males, but, again, only the mitochondrial DNA of the female children will be passed on in turn.[41] For this reason, mitochondrial DNA was particularly successful in identifying the grandchildren of Argentinean women who had lost sons and daughters during the terrible political unrest of the 1970s and 1980s.[42]

Mitochondrial DNA offers a high degree of accuracy as well. As one source explains:

> There are only four nucleotides that make up all mitochondrial DNA: cytosine, adenine, thymine, and guanine, known as CATG for short. In mitochondrial DNA there are 16,569 base pairs of nucleotides, arranged in a ring. A computer printout of a DNA sequence looks like a diabolically complex code, based on just four letters in neat columns, repeated again and again in slightly varying order for page after page. Luckily we do not have to scrutinize all 16,569 pairs. We can focus on certain "hyper-variable regions," made up of a total of just 608 base pairs. Computers are of great help in matching up the hyper-variable regions. When the results for the hyper-variable regions in two DNA samples match up at these crucial checkpoints, you can be virtually certain they are the same.[43]

With its longevity and accuracy, mitochondrial DNA is an ideal technique for the forensic anthropologist in establishing the identity of skeletal remains. Even the process of obtaining exemplar samples is easy. A ready source of such DNA for comparison with questioned remains is hair from the hairbrush or comb, or even bedding or a plumbing fixture trap that is associated with a particular missing person. As we shall see presently, DNA testing was successful in resolving the question of the identities of a group of skeletons unearthed in 1979 and believed to be the remains of the Romanovs—the royal family executed in Russia in 1918.

CASE STUDY: THE ASSASSINATION OF THE ROMANOVS

As an outcome of World War I, military defeats coupled with high casualties resulted in widespread discontent in the Russian populace directed at their tsar, Nicholas II, who was forced to abdicate in March 1917. When a provisional government failed to end the war, there were riots, massive desertions, and factional fighting. Finally, the Bolsheviks under Lenin overthrew the moderate socialist government in a violent coup. Although they ended the war with Germany, their brutal suppression of all opposition provoked civil war. The Bolsheviks and their supporters were allied with the Red Army; in opposition to them were assorted anarchists, monarchists (supporters of the royal family, the Romanovs), and ethnic groups (Poles, Georgians, and Ukrainians).

Finally, Lenin resolved to assassinate Tsar Nicholas and his family, which consisted of his wife, Alexandra Feodorovna, the tsarina of Russia (formerly Princess Alix of Hesse and the Rhine); their three daughters, Olga, Tatiana Maria, and Anastasia; and their thirteen-year-old son, Alexis Nicolaievich, the heir-apparent, or tsarevich. Alexis was a hemophiliac and was attended by a personal physician. The family also had three domestic servants. On the night of July 16, 1918, in Ekaterinburg, Siberia, the royal Romanov family was staying at a mansion that had been requisitioned from an engineer. They were told that they were to be evacuated due to nearby fighting. About midnight, they were summoned to the cellar where they were awaited by a death squad headed by Commander Jacob Yurovsky. As the shocked family and their servants struggled to comprehend what was happening, the decree of execution was read aloud. "What? What?" the tsar shouted. "Aren't we going to get out of here after all?"[44] The commander's reply was to draw his pistol and fire directly into the tsar's face, sending a bullet through his brain. He spun around and fell to the cellar floor. Another soldier, Peter Ermakov—known as "Comrade Mauser"—killed the tsarina. From a distance of only six feet, he shot her in the mouth with his Mauser, and she fell dead. Commander Yurovsky shot the tsarevich, knocking him out of his chair. While he lay groaning on the floor, a hail of bullets struck the others. The family doctor, Sergei Botkin, attempted to turn away but was hit in the neck. The youngest girls, Anastasia and Maria, fell beside the physician. "Comrade Mauser" turned to the cook, who was cowering in a corner, and shot him once in the body and again in the head. Someone else shot the tsar's footman. The maid, Anna, escaped the first onslaught, having hidden behind two pillows that had been filled with jewels. She was dispatched with a bayonet thrust through the throat. The tsarevich, who was just two weeks short of his fourteenth birthday, was still groaning and writhing on the floor. Yurovsky fired two more pistol shots at his head, and he lay dead. Anastasia was also discovered to be still alive. As with her sisters, jewels sewn into her corset had deflected some of the bullets and prolonged her agony. When a guard turned her onto her back, she shrieked. The guard clubbed her to death with the butt of his rifle.[45]

After some twenty minutes, the corpses were loaded into a Fiat truck. "Comrade Mauser" had spent a day searching the area around Ekaterinburg for a place to dispose of the bodies. He had chosen an abandoned mine shaft some twelve miles from town, and Yurovsky had given his approval. "By the light of the lamps," Ermakov recalled years later as he was on his death bed, "we stripped the corpses of their clothes. Found a lot of diamonds sewn in the Tsarina's bodice—and more necklaces, gold crosses and a lot of other such things on the girls. These were sent to Moscow along with everything else."[46] The revolutionaries burned the clothes and took the bodies to the mine and dumped them down the shaft.

To his chagrin, Yurovsky discovered that word of the bodies' whereabouts had become widespread, and the bodies had to be relocated. The following night, by torchlight, ropes were tied around the corpses, and they were hauled out of the mine, taken to another site, and there doused with hydrochloric acid. Yurovsky burned two of the bodies, that of the tsarevich and the body of a woman whose identity he was confused about, first saying it was the tsarina but later concluding it was her maid. (As we shall see presently, he was apparently wrong on both counts.) Yurovsky had meant to burn all the bodies but had underestimated the time and fuel necessary.

Just eight days after the brutal slayings, the Russian White Armies reconquered the Siberian town, and investigators found the scene of the crime. The house was a shambles, and the basement room—although recently scrubbed—had walls that were pockmarked with bullet holes and gouges from bayonets. Local rumors led the investigators to an abandoned mine where they found ashes and debris from the burning of clothes. Amid the debris they found belt buckles and parts of military caps they thought had belonged to the tsar and tsarevich; an Ulm Cross (a military decoration); the tsarina's eyeglass case; buttons, corset hooks, and shoe buckles from the grand duchesses; and other items including an eyeglass lens, three small religious icons, an emerald cross, a pearl earring, topaz beads, and what was identified as the upper dental plate of Dr. Botkin.[47]

The investigators did not learn the fate of the corpses. That remained for the decade of the 1980s when the Soviet Union under President Mikhail Gorbachev launched a period of "openness" and declassified thousands of previously secret documents. One of these was the "Yurovsky Note," an account of the assassination by Commander Jacob Yurovsky.[48]

Years earlier, another part of the drama had begun to unfold outside the Soviet empire. In 1920, following a suicide attempt in which she was pulled from a Berlin canal, an amnesiac victim began to have strange recollections. Although she could not speak Russian, a fact attributed to her "amnesia," she claimed to be the tsar's youngest daughter, Anastasia. According to her tale, two Red Army soldiers had discovered her alive and had diverted her from the Fiat truck to a cart and eventual safety. She bore a strong resemblance to the grand duchess, and a former lady-in-waiting to the tsarina even provided a positive identification. Over the years, so did certain marginal members of the royal family, although others—including Anastasia's godmother and her Swiss tutor—rejected her claim. As early as 1927, a private investigator concluded that she was actually a Polish peasant who had vanished just three days before the mystery woman appeared in the canal. "Anastasia" took the name Anna Anderson for privacy but maintained her claim to royalty until her death in Charlottesville, Virginia, in 1984.[49]

In the increasing new "openness" of the Soviet Union, a filmmaker for the Interior Ministry, Gely Ryabov, became obsessed with finding the lost graves of the Romanovs. Tracking down the children of Commander Yurovsky and obtaining a copy of the "Yurovsky Note" and also aided by a local historian and another man, Ryabov eventually uncovered a layer of logs beneath which was a pile of blackened, greenish bones. He felt certain the bones were those of the royal family. This occurred in 1979, but Ryabov waited a decade until he felt secure enough to make his discovery public.

In 1992, the Russians asked American forensic experts for technical assistance in identifying the remains, and famed forensic anthropologist Dr. William R. Maples assembled a team of distinguished experts including Dr. Lowell Levine and Dr. Michael Baden; a hair and fiber microscopist, Cathryn Oakes; and Maples's wife, Margaret, a media specialist who would assist in documenting the investigation. Later two Florida medical examiners would also accompany the team on trips to Ekaterinburg.[50]

In all, there were nine skeletons, five female and four male, rather than the expected eleven that had been executed in 1918 (seven members of the Romanov family plus cook, footman, maid, and physician).

Each of the nine skulls had a badly fractured face, making reconstruction of the features "risky or impossible," according to Maples, "but it also conformed to the accounts of the assassinations: that the faces of the victims were smashed in with rifle butts to render them unrecognizable."[51]

Other facts added to the realization that this was the long-sought grave. In addition to the bones, fourteen bullets were recovered from the mass burial where they apparently had fallen loose as the remains decomposed. Also found were bits of rope and a broken jar that once had held sulfuric acid; all of the remains showed evidence of acid corrosion. Three skulls had through-and-through gunshot wounds; another skeleton had a stab wound in the breastbone that was consistent with a bayonet wound. Tentative identifications of the bodies were as follows (the numbers being those originally assigned by the Russians):

1. The pelvis showed it was a fully grown female. The ankle joints "showed an extension of the joint surfaces, as if the woman had spent many hours crouching or kneeling, perhaps while she was scrubbing floors or doing menial work." Tentative identification: the tsarina's maid, Anna Demidova.[52]

2. The remains were that of a mature man. The torso was still intact, due to the adipocere (a waxy substance that forms on dead tissue due to excessive moisture). This permitted two bullets to be recovered, one from the pelvic area, another from the vertebra. The skull exhibited another gunshot wound. It lacked upper teeth, consistent with having had an upper dental plate. Tentative identification: Dr. Sergei Botkin, who had attended the Tsarevich Alexi.[53]

3. The skeleton was that of a female in her early twenties. The shape of the head with its bulging forehead agreed closely with photographs of Olga. The skull exhibited a gunshot wound with a trajectory that began under the jaw and exited through the frontal bone, consistent with being fired at while lying on the floor. Tentative identification: Grand Duchess Olga.[54]

4. The skeleton was that of a relatively short, middle-aged man. In part, the evidence showed the following: "The skull had a very broad, flat palate that is consistent with the mouth shape of the Tsar in photographs taken before he grew his beard. It had a jutting brow line, and so did the Tsar: the curving, protruding supraorbital bones are consistent with photographs of Nicholas taken during his life." In addition, "The hipbones showed the characteristic wear and deformation produced by many hours on horseback, and we know the Tsar was an ardent horseman." Tentative identification: Nicholas II, Tsar of Russia.[55]

5. The skeletal remains were those of a woman in her late teens or possibly early twenties—the youngest of the five female skeletons, based on "the fact that the root tips of her third molars were incomplete." Also, "Her sacrum, in the back of her pelvis, was not completely developed. Her limb bones showed that growth had only recently ended." A bullet was found near this body, in a lump of adipocere. Tentative identification: Grand Duchess Maria, who was nineteen at the time of the executions.[56]

6. This skeleton belonged to a fully grown young woman. She had a mature sacrum and pelvic rim, indicating a minimum age of eighteen, while her mature collarbone made her at least twenty. Her height was estimated at 65.6 inches, right between the other two young females." She had been shot in the back of the head. Tentative identification: Grand Duchess Tatiana, who was just past twenty-one when murdered.[57]

7. The skeleton of this body belonged to "an older woman" with "amazing and exquisite dental work," including two crowns made of platinum, others beautifully made of porcelain, and "wonderfully wrought gold fillings." The rib cage appeared to show evidence of bayonet thrusts but was too deteriorated for a definite determination. Tentative identification: the Tsarina Alexandra.[58]

8. This skeleton was "very fragmentary" and "grievously damaged by acid." Nevertheless, it clearly belonged to "an adult male in his forties or fifties" who was not very big. The area of the eyebrows was notably flat, indicating that the man had had a "flattened profile" (unlike the tsar). Also, one ulna exhibited an old fracture. Tentative identification (largely by "a process of elimination"): the Romanovs' cook, Iran Kharitonov.[59]

9. This skeleton was that of "a big, heavy-boned man over six feet tall, who was beginning to show evidence of aging." Overall, "the robust size of the skeleton agrees well with the descriptions we have of the footman." Tentative identification: the tsar's footman, Alexei Trupp.[60]

Having identified Olga, Maria, and Tatiana, Maples asks:

Where was Anastasia? None of these three young female skeletons was young enough to be Anastasia, who was seventeen years and one month old the night of the shootings. Our Russian hosts believed that body No. 6, the midmost of the three young females, was the long-lost Anastasia. Alas! We had to disagree, based on the growth patterns of the teeth, pelvises, sacra and long limbs of the three skeletons before us. The Russians had labored manfully over Body No. 6, attempting to restore its facial bones with generous dollops of glue, stretched across wide gaps. They had been forced to estimate over and over again, while reassembling these fragments, almost none of which were touching each other in the reconstruction. It was a remarkable and ingenious exercise, but it was too fanciful for me to buy: Anastasia was not in this room.[61]

Could it be that Anna Anderson was Anastasia after all? Or was hers the body that had been burned by Yurovsky and misidentified by him as the maid, Anna?

In 1993, what Maples calls "the story of the royal bones" was dramatically developed when mitochondrial DNA tests were done. Since both the tsarina and the husband of Britain's Queen Elizabeth II, Prince Philip, were maternal-line descendants of Queen Victoria, experts obtained a blood sample from the prince and extracted its mitochondrial DNA. The Russians then supplied a sample of the Romanov bones from which a small amount of DNA was extracted (by the technique of PCR, or polymerase chain reaction, discussed in chapter 8) and grown in a culture to expand its quantity. The result was a match with a probability of almost 99 percent. Maples concluded that "Taken in conjunction with the compelling physical skeletal evidence, the results are clear and unequivocal."[62]

Meanwhile, "Anna Anderson" was dead, having maintained that she was Grand Duchess Anastasia to the end. Her body was cremated, precluding any DNA testing. Samples of her hair were reportedly available, but they were cut samples, missing the roots that were desirable for DNA tests. However, prior to her death Anna Anderson had had an operation, and the hospital had retained a tissue sample. DNA from this tissue was compared first with a Romanov bone sample. DNA expert Dr. Peter Gill concluded, "The sample that we got from the tissue did not match the DNA profile which we would expect to have found from the Grand Duchess Anastasia."[63]

Next, the question of Anna's alleged real identity as a Polish factory worker, Franzisca Schanzkowska, was considered. A maternal grand-nephew of Franzisca, Carl Maucher, was located in Germany, and he provided a blood sample for DNA comparison. The result was what Dr. Gill termed "a positive match."[64] This proves that Anna Anderson was not the lost Anastasia but an impostor, the missing Polish peasant Franzisca Schanzkowska.

In 1998, eighty years after the assassinations of the Romanovs, their funeral was held in St. Petersburg, Russia. Russian officials concluded that the remains were those of the tsar, the tsarina, three of their children (excluding, they determined, Alexei and Marie), and four members of their retinue. The small coffins containing the bones were draped with the imperial flag, although many in attendance continued to dispute the identifications despite the positive DNA results.[65]

NOTES

1. The late Dr. David Wolf, comment to Joe Nickell, circa 1983.

2. William R. Maples, *Dead Men Do Tell Tales: The Strange and Fascinating Cases of a Forensic Anthropologist* (New York: Doubleday, 1994), 3.

3. Ibid., 42.

4. Ibid., 61-74, 124, 149, 151.

5. Ibid., 25, 27, 123-24, 145, 149-51.

6. Ibid., 232.

7. Daniel J. Hughes, *Homicide: Investigative Techniques* (Springfield, Ill.: Charles C. Thomas, 1974), 316-17.

8. Ibid., 317.

9. Maples, *Dead Men*, 47-48.

10. Ibid.

11. For arsenic's preservative effects, see Thomas A. Gonzales et. al., *Legal Medicine: Pathology and Toxicology*, 2nd ed. (New York: Appleton-Century-Crofts, 1954), 741; for those on antimony see Charles E. O'Hara, *Fundamentals of Criminal Investigation*, 3d ed. (Springfield, Ill.: Charles C. Thomas, 1973), 532. For mummification, etc., see Joshua A. Perper, "Time of Death and Changes after Death," in Werner U. Spitz, ed., *Spitz and Fisher's Medicolegal Investigation of Death*, 3d ed. (Springfield, Ill.: Charles C. Thomas, 1993), 32, 36-38.

12. Charles R. Swanson Jr., Neil C. Chamelin, and Leonard Territo, *Criminal Investigation*, 4th ed. (New York: McGraw-Hill, 1988), 261-62.

13. Maples, *Dead Men*, 25-26.

14. Vernon J. Geberth, *Practical Homicide Investigation*, 2nd ed. (Boca Raton, Fla.: CRC, 1993), 207-9.

15. Marcella Farinelli Fierro, "Identification of Human Remains," in Spitz, *Medicolegal*, 124.

16. Geberth, *Practical Homicide Investigation*, 209.

17. Fierro, "Identification," 97. (Information for the table was taken from this source.)

18. Ibid.

19. Geberth, *Practical Homicide Investigation*, 210.

20. Fierro, "Identification," 97.

21. Ibid., 100-1; Geberth, *Practical Homicide Investigation*, 210.

22. Geberth, *Practical Homicide Investigation*, 210.

23. Maples, *Dead Men*, 28-29.

24. Ibid., 217-18.

25. Geberth, *Practical Homicide Investigation*, 210.

26. Irvin M. Sopher, "Forensic Odontology," in Spitz, *Medicolegal*, 118-21.

27. Geberth, *Practical Homicide Investigation*, 199.

28. Sopher, "Forensic Odontology," 123.

29. Ibid.

30. Ibid., 119.

31. Ibid., 120-21.

32. Ibid., 127-32; Geberth, *Practical Homicide Investigation*, 205.

33. Colin Evans, *The Casebook of Forensic Detection: How Science Solved 100 of the World's Most Baffling Crimes* (New York: John Wiley & Sons, 1996), 151-54. See also Richard W. Larsen, *Bundy: The Deliberate Stranger* (New York: Pocket Books, 1986), 288-90, 311-13.

34. Maples, *Dead Men*, 55-56.

35. Clyde Collins Snow in Geberth, *Practical Homicide Investigation*, 216-17.

36. Snow, Ibid., 217.

37. Fierro, "Identification," 79.

38. Ibid.

39. Gerald L. Posner and John Ware, *Mengele: The Complete Story* (New York: McGraw-Hill, 1986), 320, 325.

40. Maples, *Dead Men*, 265.

41. Ibid.

42. Fierro, "Identification," 111.

43. Maples, *Dead Men*, 266.

44. Except as noted, information for this case study was taken from ibid., 238-68, and Evans, *The Casebook of Forensic Detection*, 58-60.

45. The details were provided in part by Peter Ermakov ("Comrade Mauser") on his deathbed to Richard Halliburton for his 1935 book, *Seven League Boots*, and also by other participants, including the commander, Jacob Yurovsky, all cited or quoted in Maples, *Dead Men*, 238-51.

46. Ermakov, in Maples, *Dead Men*, 243-44.

47. Maples, *Dead Men*, 244-45.

48. See n. 46.

49. Evans, *Casebook*, 59; Gordon Stein, ed., *Encyclopedia of Hoaxes* (Detroit: Gale, 1993), 105.

50. Maples, *Dead Men*, 250-52; Evans, *Casebook*, 58.

51. Maples, ibid., 252-53.

52. Ibid., 254.

53. Ibid., 254-55.

54. Ibid., 255.

55. Ibid., 259.

56. Ibid., 256.

57. Ibid.

58. Ibid., 258-59.

59. Ibid., 257-58, 260-61.

60. Ibid., 258.

61. Ibid., 256-57.

62. Ibid., 266-67.

63. Peter Gill, quoted on "Anastasia: Dead or Alive," *Nova*, Oct. 10, 1995.

64. Ibid.

65. Paul Quinn–Judge, "Final Resting Place," *Time*, July 27, 1998.

RECOMMENDED READING

Holland, Mitchell M., et al. "Mitochondrial DNA Sequence Analysis of Human Remains." *Crime Laboratory Digest*, vol. 22, no. 4 (Oct. 1995). Forensic discussion of mitochondrial DNA, which is known (in addition to chromosomal DNA) as "the other human genome." Explains its suitability to identification of human remains, and includes two case studies and extensive references.

Iserson, Kenneth V. *Death to Dust: What Happens to Dead Bodies?* Tucson, Ariz.: Galen Press, 1994. A comprehensive discussion of all aspects of death, including autopsies, burials, funerals, cryogenic preservation, cremation, grave robbing, forensic pathology and anthropology, epitaphs, mummification, cannibalism, and so on and on.

Joyce, Christopher, and Eric Stover. *Witnesses from the Grave: The Stories Bones Tell.* New York: Ballantine Books, 1991. Case studies of forensic anthropologist Clyde Snow, including examination of the remains of Nazi doctor Joseph Mengele, the cavalrymen who died at Custer's Last Stand, and the victims of serial killer John Wayne Gacy.

Kurth, Peter. *Anastasia: The Riddle of Anna Anderson*, updated edition. Boston: Little, Brown, 1986. Stubborn defense of Anna Anderson's claim to be the lost Russian Grand Duchess Anastasia, updated with forensic anthropological information on the discovered Romanov remains—missing Anastasia's—but without the final DNA proof that Anna Anderson was an impostor. Recommended for its discussion of this important controversy.

Maples, William R. *Dead Men Do Tell Tales.* New York: Doubleday, 1994. Case studies of forensic anthropologist Dr. William R. Maples, including examinations of the remains of Francisco Pizarro, the conquistador; President Zachary Taylor; and the assassinated royal family of Russian Tsar Nicholas II.

Ubelaker, Douglas, and Henry Scammell. *Bones: A Forensic Detective's Casebook.* New York: Edward Burlingame Books (HarperCollins), 1992. Case studies of noted forensic anthropologists, including Dr. Douglas Ubelaker, curator of Anthropology at the Smithsonian Institution and a regular consultant to the FBI laboratory. Wide-ranging and well-illustrated.

AFTERWORD

In the preceding chapters, we have attempted to portray the various forensic sciences as they typically are carried out today at crime scenes and in laboratories, and we have endeavored to provide perspective by discussing the pioneers and techniques of the past. We have seen how the past century has placed crime detection on a scientific foundation.

There is an old saying that the more things change, the more they stay the same, and this is certainly true of the forensic sciences. Although the practitioners of a century ago would scarcely recognize the present-day laboratory with its computerized AFIS system, DNA typing, and gas chromatography, still the basic elements of fingerprinting, bloodstains, and suspected poisons remain. They are among the all-too-constant elements of the criminal domain. Also, according to Richard Saferstein, "Although it is true that over the past two decades, owing to developments in analytical instrumentation, there have been dramatic changes in the tools and techniques available to the forensic analyst, human involvement is still required to interpret and weigh the significance of data emanating from these machines."[1] If our forebears would be astonished at the current technology, we can expect future developments to be no less amazing. Yet it is not as easy to see the future, which must be imagined, as to view the past, which readily offers its images and texts for us to peruse.

It does seem likely that fingerprinting and other forensic standbys will continue well into the future. Common sense shows why. As David Fisher explains in his *Hard Evidence*: "There is a misconception that DNA eventually will replace just about every other sci-crime technology, particularly fingerprinting. This isn't true. DNA provides evidence that has never before been available in violent crimes, but it's complementary to prints. DNA is rarely found on paper, for example, but prints found on paper provide valuable evidence in a variety of crimes from money laundering to bank robberies to kidnapping."[2]

Certain trends may also continue, such as that toward specialization, which has been held to be "irreversible."[3] At the same time, if it is not desirable for the modern criminalist to be a jack-of-all-trades, he or she must at least have a generalist's knowledge of what is possible in other fields. Otherwise, too much compartmentalization may take place, with a resulting loss of opportunity for cooperation between specialists.[4]

Above all, the trend toward scientific progress seems sure to continue. Just as the past has yielded such remarkable developments as fingerprinting and DNA profiling, together with computer technology, other breakthroughs no doubt await discovery. It is the least we can expect in the unrelenting fight against crime.

NOTES

1. Richard Saferstein, "Forensic Science: Winds of Change," in Samuel M. Gerber, ed., *Chemistry and Crime* (N.P.: American Chemical Society, 1983), 41.

2. David Fisher, *Hard Evidence* (New York: Dell, 1995), 190.

3. Saferstein, "Forensic Science," 41.

4. Ibid.

INDEX

Note: Italicized page numbers indicate illustrations.